cosmology

Jim Breithaupt

TEACH YOURSELF BOOKS

For UK orders: please contact Bookpoint Ltd, 39 Milton Park, Abingdon, Oxon
OX14 4TD. Telephone: (44) 01235 400414, Fax: (44) 01235 400454. Lines are open
from 9.00–6.00, Monday to Saturday, with a 24-hour message answering service.
Email address: orders@bookpoint.co.uk

For USA & Canada orders: please contact NTC/Contemporary Publishing, 4255 West
Touhy Avenue, Lincolnwood, Illinois 60646–1975, USA. Telephone: (847) 679 5500,
Fax: (847) 679 2494.

Long renowned as the authoritative source for self-guided learning – with more than
40 million copies sold worldwide – the *Teach Yourself* series includes over 200 titles in
the fields of languages, crafts, hobbies and other leisure activities.

A catalogue entry for this title is available from The British Library.

Library of Congress Catalog Card Number: On file

First published in UK 1999 by Hodder Headline Plc, 338 Euston Road, London NW1 3BH.

First published in US 1999 by NTC/Contemporary Publishing, 4255 West Touhy Avenue,
Lincolnwood (Chicago), Illinois 60646–1975, USA.

The 'Teach Yourself' name and logo are registered trade marks of Hodder & Stoughton Ltd
in the UK.

Typeset by Transet Limited, Coventry, England.
Printed in Great Britain for Hodder & Stoughton Educational, a division of Hodder
Headline Plc, 338 Euston Road, London NW1 3BH by Cox & Wyman Ltd, Reading,
Berkshire.

Impression number   10 9 8 7 6 5 4 3 2 1
Year   2005 2004 2003 2002 2001 2000 1999

CONTENTS

Preface _____ v

Acknowledgements_____ vii

1 | What is Cosmology?_____ **1**
Questions from long ago _____ 1
What do cosmologists do? _____ 2

2 | Beyond Imagination _____ **5**
Seeing is believing _____ 5
Sizes and scales _____ 10

3 | The Oldest Science _____ **16**
About the Celestial Sphere_____ 16
The first cosmologists _____ 24

4 | The Scientific Revolution _____ **27**
Scientific controversies _____ 27
Newton's Universe _____ 38
Laws and theories_____ 41

5 | Strange Ideas_____ **45**
Space and time _____ 45
The cosmic speed limit_____ 52
Mass and energy_____ 56

6 | Giants and Dwarfs _____ **63**
Making measurements _____ 63
Contrasts and comparisons_____ 72
Star spectra_____ 76

7 | The Life of a Star_____ **85**
Binary stars_____ 85
Inside a star _____ 91
The life cycle of a star _____ 96
Star lifetimes _____ 105

8 | Towards the Edge of the Universe _____ **108**
Mapping the Milky Way_____ 108
Beyond the Milky Way_____ 121
Hubble's great discovery _____ 129

9 | Black Holes _____ **136**
Gravity and light _____ 136
About black holes_____ 143
Ideas beyond imagination _____ 146
Testing times _____ 149

10 | The Big Bang _____ **154**
Before the Big Bang theory _____ 154
The beginning of time _____ 157
Evidence for the Big Bang theory _____ 168

11 | Fundamental Forces _____ **173**
Forces and fields _____ 173
Smaller and smaller _____ 181

12 | The Early Universe _____ **192**
Images from the past _____ 192
Inflation at work_____ 199
The history of the early Universe_____ 205

13 | Into the Future _____ **214**
The dark age of the Universe _____ 214
Factors for the future _____ 220
An uncertain future _____ 226

Glossary _____ **231**
Appendices _____ **239**
1 Proof of Kepler's 3rd Law (Chapter 4) _____ 239
2 More about special relativity (Chapter 5) _____ 240
3 Taking account of gravity (Chapter 10) _____ 242
4 Notes on Einstein's General Theory of Relativity
  (Chapters 9 and 13) _____ 243
5 Calculation of the critical density of the Universe
  (Chapter 13) _____ 245
6 Spreadsheet for 'Out of chaos' (Chapter 13) _____ 246

Index _____ **247**

PREFACE

Cosmology has always been at the forefront of science from ancient times to the present day. The search for knowledge and understanding about the Universe, its origin and its future is a very active branch of science, perhaps more so now because of recent discoveries and new technology. Experiments that probe matter on the smallest possible scale have led to conclusions and theories that cosmologists are using to explain astronomical observations on the largest possible scale. The primary aim of this book is to provide for the non-specialist an account of cosmology at present, including the links with science on the smallest scale deep inside the atom as well as developments and discoveries on the largest scale in the remote Universe of the distant past. In addition to looking in detail at present-day cosmology, the key role of modern science in cosmology becomes evident when we look at the development of cosmology before the Scientific Age which began about four centuries ago. New technology in the form of the telescope ushered in the Scientific Age when Galileo made the first-ever detailed observations of the lunar surface and discovered the moons of Jupiter. Ever since, new astronomical discoveries have resulted from new instrumentation, most recently in the shape of the Hubble Space Telescope. In recent decades, computing power on a scale unbelievable half a century ago has added to the technological armoury of modern cosmology.

The book is written for the beginner with a general interest in science who is curious to find out not only about the Big Bang theory of the origin of the Universe, but also about other ideas of the Universe and the influence of these ideas on human thought and behaviour over thousands of years. The scientific principles that underpin present knowledge are explained in detail, and the crucial role of experimental science is described. Activities which mostly require no more than everyday materials are provided at intervals to reinforce understanding.

The ideas of cosmology are as fundamental at the present time as they were in ancient times. Fortunately, modern science limits the scope of such ideas, preventing rigid doctrines from being imposed. I hope this book conveys the astonishing discoveries and theories of cosmology as well as its profound significance for our future.

ACKNOWLEDGEMENTS

I would like to thank my family and my colleagues at Wigan and Leigh College for their support in the preparation of this book, particularly my wife for secretarial support and continued encouragement. I am also grateful to the publishing team at Hodder and Stoughton, in particular Joanne Osborn and Jill Birch who initiated and oversaw the project.

I would also like to thank the following individuals and organizations for the use of their illustrations and photographs in the book.

Black and white photo illustrations:
Figure 3.1 Star trails: John Sanford/Science Photo Library
Figure 8.5(a) Sample evidence for Hubble's law: © Carnegie Institution of Washington
Figure 8.5(b) Photo of Edwin Hubble: Science Photo Library

Colour plates
Plate 1 Comet Hale Bopp: Frank Zullo/Science Photo Library
Plate 2 The Crab Nebula: Jeff Hester and Paul Scowen, Arizona State University/Science Photo Library
Plate 3 The Milky Way: Dr Fred Espenak/Science Photo Library
Plate 4 M3, a globular cluster: NOAO/Science Photo Library
Plate 5 The Andromeda galaxy, M31: Tony Hallas/Science Photo Library
Plate 6 Gravitational lensing: Space Telescope Science Institute/ NASA/Science Photo Library
Plate 7 M87: Space Telescope Science Institute/NASA/Science Photo Library
Plate 8 M82, a galactic survivor: Chris Butler/Science Photo Library
Plate 9 The Large Hadron Collider: David Parker and Julian Baum/Science Photo Library
Plate 10 COBE's microwave map of the Universe: NASA/Science Photo Library

1 | WHAT IS COSMOLOGY?

Before setting out on the journey from some of the ancient ideas about the Universe to our present knowledge, this chapter surveys the terrain covered by modern cosmology. This survey is intended to prevent unnecessary wandering into areas beyond the realm of modern cosmology. The questions that have shaped modern cosmology are set out in this chapter in addition to the questions that are beyond modern cosmology.

Questions from long ago

The simplest questions are often the most difficult to answer. Simple questions often raise more questions instead of answers. Cosmology is a scientific discipline based on simple questions such as 'Is the Universe finite?' Such simple questions are big questions in the sense that they are about nature on the largest possible scale. Cosmology has been at the forefront of human endeavour ever since humans discovered how to record their thoughts and activities. Even before recorded history, humans probably wanted to know the answers to the simple questions which cosmology is based on. After thousands of years, at the start of a new millennium, some of the answers are being discovered. However, unlike many branches of science where asking the right questions can be the key to solving a difficult problem, the big questions of cosmology aren't very difficult to raise but they have turned out to be very difficult to answer. Before considering what scientists think these questions are, pause for a few minutes and make up your own list of what you, the reader, consider to be the big questions of cosmology.

So what are the big questions of the science of modern cosmology? Compare the list on the next page with your own list, bearing in mind that big questions tend to generate more questions.

Is the Universe finite or infinite? If the Universe is finite, how big is it?

Was there a beginning for the Universe? If so, how long ago did the Universe begin?

Will the Universe end? If so, how will it end?

Is the Universe changing? If so, how did it reach its present state and how will it change in future?

Where do we fit in?

Are the laws of science universal?

What about the questions ruled out by modern cosmologists? If you have any of the questions below on your own list, don't expect answers from modern cosmologists. The questions are not unreasonable questions to ask but they form no part of modern cosmology.

Who created the Universe?

Why was the Universe created?

Why do we exist?

If any of the above questions are on your list, cross them off and forget about them as you read this book. However, don't be afraid to ask these questions elsewhere but equally don't expect them to be part of this or any other book on cosmology.

What do cosmologists do?

We live in a scientific age in which the amount of new knowledge discovered in the past few decades far exceeds all the previous knowledge known. Cosmology is a very active branch of science, drawing upon the principles of mathematics and physics as applied to astronomy, chemistry and electronics. All these subjects underpin our scientific age. It began several centuries ago, when scientists like Galileo Galilei realized that the road to scientific knowledge and understanding is based on the guiding rule that the theories of science must fit the facts, and not the other way round.

Before the scientific age, philosophers sifted through facts and observations and rejected any facts that did not fit their theories. The astronomers of Ancient Greece put the Earth firmly at the centre of their model of the Universe. They imagined the stars were attached to an invisible 'celestial sphere' which was supposed to revolve round the Earth. This neat model explains why stars move across the night sky but more invisible spheres were needed to explain the motion of the planets. The model became very complicated as astronomers and philosophers attempted to make it explain their observations.

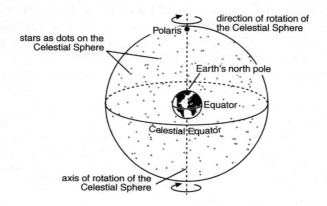

Figure 1.1   The Celestial Sphere

In the 16th century, Nicolas Copernicus, a Polish monk, put forward the theory that planets and the Earth orbit the Sun. This theory was in direct conflict with the teachings of the Church, which held that the Earth was at the centre of the Universe and that Heaven was beyond the Celestial Sphere. Copernicus's heliocentric theory of the Universe never obtained any support from the Church during his lifetime. He died in 1543, unaware that he had sown the seeds for a revolution initiated by Galileo that laid the foundations for the scientific age. Galileo is the 'architect' of modern science because he invented a telescope powerful enough to study the planets and he used his observations to demonstrate that the Copernican model is correct. We will take a more detailed look at the work of Copernicus and Galileo later in this book.

Cosmologists nowadays have equipment and resources which not even Galileo could have imagined. Nevertheless, the scientific method developed by Galileo and others continues to be the guiding rule for scientists of all disciplines. Cosmologists develop and use extremely sophisticated scientific instruments and high-power telescopes to observe the most distant objects known. They use very powerful computers to analyse their observations. They work on very complicated mathematical theories to make predictions to test their theories. They study the nature of matter on a scale which is almost unimaginably small as they strive to understand the origins of the Universe. In short, cosmology in the scientific age is a highly specialized branch of science which has the primary aim of answering fundamental questions about the Universe. Although the work of cosmologists is directed towards the primary aim of cosmology, many important technological developments in optics, electronics, space technology and computing have resulted from the work of cosmologists.

Cosmologists are perhaps more aware than most scientists of the need to make their work understandable to non-scientists as well as to scientists in other disciplines. The big questions of cosmology are of great interest to many people from all walks of life. The answers to some of the big questions are now known and have turned out to be not too complicated, although perhaps surprising in view of what many people would expect as answers. The questions with known answers are listed below. What would you expect for an answer to each of these questions?

Is the Universe finite or infinite?
If the Universe is finite, how big is it?
Was there a beginning for the Universe?
If so, how long ago did the Universe begin?

Until about 30 years ago, many cosmologists thought that the Universe is infinite, without a beginning or an end. However, scientific discoveries since then have led to the generally accepted conclusion that the Universe is:

■ **finite**
■ **about 10 000 million years old**
■ **the result of a massive explosion from a point, the so-called Big Bang**
■ **probably still expanding.**

These conclusions are perhaps surprising because the scientific evidence is very firm, except on the last point, and leaves little doubt in the minds of scientists, including those who advocated other theories several decades ago. The science behind these answers forms a large part of this book which also looks at progress towards answering the other big questions of cosmology. More questions have arisen from the answers discovered thus far. What questions can you think of in the light of the conclusions listed above? We will consider some of these further questions in the final chapter. Now let's look in more detail at the science behind the big questions of cosmology.

2 | **BEYOND IMAGINATION**

In this chapter, we start our journey from the past to the present by describing the main features of the night sky as observed with the unaided eye and with a low-power telescope or a pair of binoculars. We then jump forward to an overview of what we now know about the distances to objects we can see in the night sky. This allows us to look at the scale of the solar system, the galaxies and the Universe. Finally, we scale the age of the Universe to one year to put the history of the human race in perspective.

Seeing is believing

A cosmic coincidence

If you didn't know otherwise, you might believe that the Sun and the Moon are about the same diameter. After all, they seem to be the same size in the sky. A small coin at arm's length will block out the Sun or the full moon. Yet the Sun is about 400 times larger than the Moon. If the Sun was represented by a football, the Moon would be about the size of a small grain of rice. The reason why they appear the same size in the sky is because the Sun is about 400 times further than the Moon from the Earth.

This cosmic coincidence is the reason why the Sun is periodically totally eclipsed by the Moon. This happens when the Moon passes between the Earth and the Sun. Where the Moon's shadow stretches out and touches the Earth, the sky becomes completely dark for a few minutes as the shadow sweeps past. If the Moon's disc was much smaller than the Sun's disc, a total solar eclipse would be impossible. In ancient times, a total solar eclipse was an event of great significance, sometimes seen as a sign of disapproval from the gods requiring ritual sacrifices to avoid doom and destruction. Now we know better as a total solar eclipse is a natural event which ancient rulers used to maintain power over their subjects!

Patterns in the sky

We see the stars in the night sky much the same as they were seen thousands of years ago. Ancient astronomers classified them according to brightness, from the brightest at first magnitude to the faintest at sixth magnitude. Thus a reference in an ancient manuscript to a star of the first magnitude would refer to one of the brightest stars in the sky. The magnitude scale is not unlike football divisions, the first division better than the second division which is better than the third division and so on. We will return to a more detailed look at the magnitude scale in Chapter 5. The brightness of most stars does not vary on a human time scale, appearing the same now as thousands of years ago. Some stars do vary in brightness periodically and these have given important clues as explained in Chapter 5 about the way stars form and evolve. The patterns formed by stars in the sky are the same now as they were thousands of years ago. This is because the stars have scarcely changed their positions relative to each other. They appear to be stationary relative to each other because they are so far away from us. Their motion has little or no noticeable effect on their positions in the sky as seen from the Earth – just the same as a vehicle on a motorway seen from an airliner several kilometres high scarcely appears to be moving. Even the nearest star to us, Proxima Centauri, is so far away it hardly changes its position relative to the more distant stars.

The constellations which we use to map the sky are patterns of stars defined by astronomers thousands of years ago in Ancient Greece. Other ancient civilizations also drew up maps of the sky in the form of constellations but it is the 88 constellations of the Greek system that we use today.

Some constellations were named after creatures because the pattern of stars was thought to resemble the creature. For example, the stars in Cygnus were thought to resemble a swan in flight with outstretched wings and a long neck. Other constellations were related to ancient myths and legends, frequently linked to ancient Gods. The Northern sky includes the constellations Cassiopeia, Andromeda, Cetus and Perseus which were linked together by the Ancient Greeks in a legend where Andromeda, the beautiful daughter of a boastful mother, Cassiopeia, was chained to a rock by a river to await an unpleasant fate when the sea-monster Cetus arrived. Fortunately, Andromeda is rescued by her hero Perseus. Clearly, the ancient astronomers linked their observations closely to their religious and cultural world. What would they have made of the idea of a Universe created in a massive explosion?

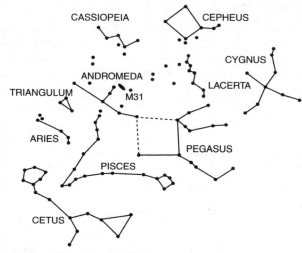

Figure 2.1   The constellations

ACTIVITY

Observe the night sky and see if you can make out the constellations in that part of the sky. If you are in the Northern hemisphere, you should be able to see Ursa Major, sometimes referred to as the Big Dipper or the Plough. Two of the stars in the Big Dipper lie along a line that passes through the Pole Star which is directly over the Earth's North Pole and therefore always due North. Continue from the Pole Star in the opposite direction to Ursa Major to locate Cassiopeia which appears like a giant W in the sky.

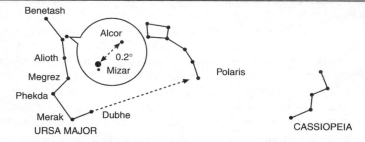

Figure 2.2   Finding the Pole Star

Wandering stars

The nine planets that orbit the Sun are (in order of increasing distance from the Sun) Mercury, Venus, Earth, Mars, Jupiter, Saturn, Uranus, Neptune and Pluto. The planets are visible from Earth because they reflect sunlight. They do not emit light as the stars do. The outer three planets are invisible from Earth without the aid of a telescope and therefore were not observed by ancient astronomers. In fact, Pluto was only discovered in 1930.

The planets move through the constellations following the same path across the sky as the Sun. Like the stars, they can't be seen in day time because the Sun makes the sky too bright. The path of the planets and the Sun across the sky defines the constellations of the **Zodiac**. The planets move through these constellations because they are all moving round the Sun in the same orbital plane, including the Earth.

- **Mercury** is the closest planet to the Sun and moves round fastest. Seen from Earth, it is never far from the Sun, visible in the East just before sunrise or in the West just after sunset. It is named after Mercury, the elusive winged messenger of the Gods of Ancient Rome.

- **Venus**, the second closest planet to the Sun, is brighter than the brightest star at times. It can be seen for two or three hours before sunrise in the East or after sunset in the West at certain times. Its brightness at times is the reason why it was named after the Goddess of Love in Ancient Rome.

- **Mars** is the nearest outer planet to the Earth. Its reddish hue makes it easy to identify as it moves gradually through the constellations of the Zodiac, appearing due South at midnight approximately every two years (i.e. at opposition; see Chapter 2). It was named after the God of War because its redness was associated with blood.

- **The asteroids** lie mostly between the orbits of Mars and Jupiter. The asteroids consist of minor planets, several hundred kilometres across, as well as countless rocks and fragments. Some asteroids are in non-circular orbits which take them inside the orbit of Mercury, crossing the Earth's orbit at regular intervals. A catastrophic collision between an asteroid and the Earth is not impossible. The closest recorded approach of an asteroid to Earth occurred on the 22nd March 1989 when an asteroid missed Earth by just 700 000 km which is less than 1/200th of the distance from the Earth to the Sun.

- **Jupiter** lies beyond Mars and is easily visible in the night sky for several months each year. It is the largest planet and its four inner moons, first observed by Galileo, can be seen through binoculars or a low-power telescope.

■ **Saturn**, the second largest planet of the Solar System, lies beyond Jupiter and is less easy to see in the night sky. Its ring system, when it is facing the Sun, can be seen through a telescope. Like Jupiter, Uranus and Neptune, Saturn is a ball of fluid in rapid rotation.

■ **Uranus, Neptune and Pluto**, the outermost planets, are too faint to see without the aid of a telescope. Uranus and Neptune, like Jupiter and Saturn, are giant spinning balls of fluid, each accompanied by several moons. Pluto, about the same size as our own Moon, is rocky and is accompanied by its own moon.

Comets

Few objects in the night sky are as breathtaking as a bright comet. In 1997, Comet Hale Bopp returned to the inner solar system after a journey lasting several thousand years taking it far beyond Pluto and back. It was thought to have been observed by the ancient Egyptians. Halley's Comet is another prominent comet and it returns to the inner solar system once every 75 years or so. It is named after Edmund Halley who correctly predicted its return in 1759.

A new comet is named after its discoverer and since comets remain dark until they approach the Sun, comet spotting is one way of achieving a place in history. The Sun's gravity keeps the planets in almost circular orbits. A comet follows an elliptical orbit, prevented from leaving the Solar System by the Sun's gravity which slows a comet down as it moves away from the Sun and speeds it up as it moves towards the Sun. Far away from the Sun, a comet is frozen solid, dark and invisible from Earth. As it approaches the Sun, its surface is heated by the Sun, making it release glowing gas and dust. The long tail of a comet is a spectacular sight and is due to the effect of the solar wind on the glowing gas and dust released by the comet. The solar wind is a steady stream of high-energy particles and radiation emitted by the Sun in all directions. The tail of a comet always points away from the Sun because of the effect of the solar wind. See Plate 1.

Nebulae

Charles Messier was an eighteenth-century French astronomer who discovered over 100 objects in the night sky which were neither comets, stars or planets. These objects were referred to as nebulae because they are fuzzy (unlike stars which are point objects) and they do not move relative to the stars like planets and comets do. Messier discovered them in his searches for comets which he knew changed position among the stars. The fuzzy objects that did not change position among the stars were catalogued by Messier and are known as Messier objects. For example, the Crab Nebula, now thought to be the remnants of a star that exploded in the eleventh century, was the first object to be catalogued by Messier and is therefore known as M1. Messier catalogued the Andromeda galaxy as the Andromeda Nebula M31 as he did not know about galaxies.

Messier objects outside the Milky Way galaxy are themselves galaxies, each consisting of millions of millions of stars. The measurement of the distances to the galaxies is part of the story of 20th-century astronomy and it led to the conclusion that the Universe consists of countless galaxies receding from each other as a result of the Big Bang. We will look at how astronomers have arrived at this picture in later chapters in this book.

Messier's catalogue of 110 nebulae was superseded in 1888 by the New General Catalogue (NGC) of Nebulae and Clusters of stars. However, the nebulae in Messier's catalogue are still referred to as Messier objects and the associated numerical designations are still used.

Sizes and scales

The stars in a constellation form a pattern because of their relative positions although they may be at vastly differing distances from Earth. Two stars that appear to us on Earth next to each other in a constellation may actually be further apart than either is to Earth. We will look later at how the distances to stars and galaxies has been worked out.

The speed of light

The scientific unit of distance is the **metre** (m). Since this is a bit small for long-distance journeys, the kilometre (km), equal to 1000 m, is often used. For example, a trip round the world along the equator would cover a distance of about 40 000 km. This is about one tenth of the distance from the Earth to the Moon which is about 380 000 km. The Moon is not too far away at a distance equivalent to ten times round the world. A jet liner flies much more than this distance every year. Getting to the Moon is much harder than flying round the world because much more energy has to be used to overcome the Earth's gravity.

Light travels through space at a constant speed of 300 000 km s^{-1}. Light takes a little over one second therefore to travel to the Earth from the Moon. The Sun is about 150 million kilometres from the Earth. How many seconds does light take to travel from the Sun to the Earth? You can work this out for yourself by calculating how many times 300 000 km divides into 150 million km. You should get an answer of 500 seconds. To check this is the correct answer, multiply this time by the speed of light (= 500 s × 300 000 km s^{-1}) and you should obtain an answer of 150 000 000 km). In other words light takes just over 8 minutes to travel from the Sun to the Earth. The mean distance from the Sun to the Earth, defined as **1 astronomical unit** (AU), is a useful 'yardstick' to express the distances to other planets. For example, the mean distance from the Sun to Jupiter is 5.2 AU which means that Jupiter is 5.2 times further from the Sun than the Earth is.

One astronomical unit (AU) is defined as the mean distance from the Sun to the Earth

Light takes over 6 hours to reach the outermost planet Pluto from the Sun. Work out for yourself the distance from the Sun to Pluto in kilometres by multiplying the speed of light (300 000 km s^{-1}) by the time taken (6 hours in seconds). You should get an answer of 6500 million kilometres, over 40 times the distance from the Sun to the Earth. Comets like Hale Bopp that take thousands of years to return to the inner solar system go far beyond Pluto, demonstrating that the Solar System extends thousands of millions of kilometres into space.

The nearest star to the Sun is a very faint star called Proxima Centauri. Light takes 4.3 years to reach us from Proxima Centauri. This is much, much greater than the distance to Pluto. Work out for yourself how far this distance is in kilometres. You should get an answer of 40 000 000 million kilometres (= 4.3 years × 365.25 days × 24 hours × 3600 seconds × 300 000 km s^{-1}).

One light year is the distance travelled by light in exactly 1 year

Light from the furthermost galaxies takes over 8000 million years to reach us. This enormous distance can be worked out in kilometres by multiplying the speed of light by 8000 million years in seconds. Work this out for yourself. The answer is on the next page.

Distance from the Sun to the Earth = 150 million kilometres
Distance from the Sun to Pluto = 6500 million kilometres*
Distance from the Sun to Proxima Centauri = 40 000 000 million kilometres
Distance from the furthermost galaxies = 100 000 000 000 million million kilometres

* This is an average distance since Pluto's orbit is elliptical

The light year, the distance travelled by light in 1 year, is a convenient unit of distance for astronomical purposes. The nearest star is about 4 light years away and the furthermost galaxies about 8000 million light years distant. Prove for yourself that 1 light year is equal to 9.5 million million kilometres. Later on in this book, we will meet another unit of distance, the **parsec** (equal to 3.26 light years), used by astronomers in preference to the light year because it relates easily to measurement of position.

ACTIVITY

Make a scale model of the Solar System

Imagine the Sun is scaled down to the size of a football. The Earth would be the size of a pea about 3 or 4 metres away. Jupiter would be the size of an orange about 20 metres away. Use the distances given earlier to show that the distance to Pluto on this scale would be about 180 metres and to Proxima Centauri about 11 000 kilometres!

The Milky Way

The Sun is just one of millions of millions of stars in the Milky Way galaxy, a spiral galaxy which is about 100 000 light years in diameter. The Sun lies in one of the arms of the galaxy which spiral outwards from the hub of the galaxy which is a central bulge. Stars lie above and below the plane of the spiral arms in a halo where the concentration of stars is less than in the spiral arms. Dust clouds prevent light reaching us from the galactic centre. However, radio waves from the galactic centre and the spiral arms are unaffected by dust and this is why radio telescopes have been used to map out the structure of the Milky Way galaxy.

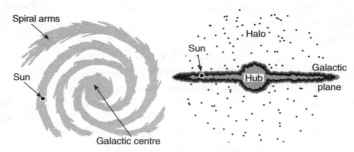

(i) In the spiral arms   **(ii) Edge-on**

Figure 2.3   The Sun in place

On a clear night, the Milky Way appears to the unaided eye as a faint irregular diffuse band of light across the sky. The centre of the galaxy lies in the direction of the constellation of Sagittarius through two spiral arms which lie between the Sun and the galactic hub. The spiral arm containing the Sun is referred to as the Orion arm after the constellation Orion the Hunter which lies in the same arm as the Sun. Orion is a dramatic sight in the Northern sky in winter, easily located by the prominent stars of Orion's belt and a sword that seems to be suspended from the belt. The red giant star Betelgeuse lies at the top left opposite the blue giant Rigel bottom right. See Figure 3.2.

Galaxies of galaxies

More than 30 nebulae catalogued by Charles Messier are now known to lie beyond the Milky Way galaxy. These extra-galactic nebulae are individual galaxies. At the end of the nineteenth century, more than 10 000 galaxies, each containing up to or more than a million million stars had been catalogued. Now astronomers reckon there are many more galaxies and they seem to be clustered together in groups with up to several thousand members.

The Milky Way galaxy is one of a cluster of at least 24 galaxies that includes the Andromeda galaxy and the Magellan Clouds which are visible from the Southern hemisphere.This cluster, referred to as the Local Group, is over two million light years across. The mean separation between the galaxies in the Local Group is about 800 000 light years. Two dozen or so apples spread out on a table would be a scaled-down model of the Local Group. One of these apples would represent the Milky Way comprising a million million stars.

Many clusters of galaxies have been observed and they are thought to be grouped in superclusters up to hundreds of millions of light years across at distances up to more than a thousand million light years away. The furthermost galaxies are thought to be over 8000 million light years away. We will return to consider these enormous distances later. For the moment, here is a recap on the sheer scale of the distances to celestial objects.

The distance from the Sun to the Earth is 8 light minutes
The distance to the nearest star is about 4 light years
The distance across the Milky Way is about 100 000 light years
The distance to the Andromeda galaxy is about 2 million light years
The distance to the furthermost galaxies is over 8000 million light years!

The one-year Universe

One year is the time taken by the Earth to go round the Sun by exactly 360°. Mars takes about 1.9 years to orbit the Sun once. Jupiter takes almost 12 years and Pluto takes about 250 years to go round the Sun once. Although there is no fundamental scientific significance in using the year as a unit of time, it is used extensively in astronomy because it is easy to relate to.

■ The Universe is thought to be over 10 000 million years old.
■ The Sun is thought to be about 5000 million years old, about half-way through its life cycle.
■ The planets are reckoned to be the same age as the Sun as they formed from the same matter.
■ The human race is little older than about 250 000 years.
■ Recorded history began about 10 000 years ago.

Imagine the age of the Universe scaled down to one year or about 500 000 minutes. On this time scale, work out for yourself how long ago the Sun and planets formed, how long ago the human race first appeared and how long ago recorded history began. The answers you should obtain are to be found on the following page.

Summary

- **A constellation** is a pattern of stars in the night sky. The night sky is mapped out in 88 constellations.
- **Planets** are visible from Earth because they reflect sunlight. The nine planets are Mercury, Venus, Earth, Mars, Jupiter, Saturn, Uranus, Neptune and Pluto.
- **Comets** are objects in elliptical orbits round the Sun, stretching far beyond the orbit of Pluto.
- **A galaxy** consists of millions of millions of stars.
- **The Sun** is one of millions of millions of stars in the Milky Way galaxy.
- **One astronomical unit** (AU) is the mean distance from the Sun to the Earth.
- **One light year** is the distance travelled by light in one year. The nearest star is about 4 light years away. The most distant galaxies are thought to be over 8000 million light years away.

In a one-year Universe,
The Sun and planets formed 6 months ago
The human race appeared 12½ minutes ago
Recorded history began 30 seconds ago.

3 | THE OLDEST SCIENCE

The big questions of cosmology have occupied the thoughts of scientists and philosophers ever since humans discovered how to record their thoughts thousands of years ago. Ancient monuments such as Stonehenge in England were undoubtedly constructed with the stars in mind, perhaps as tributes to ancient deities as well as for practical purposes such as keeping track of the exact dates of important times of the year such as mid-summer. Predicting the appearance of a planet in the night sky was an important guide to the fortunes of many an ancient ruler. Don't dismiss such activities as primitive as the astronomers of long ago were usually at the leading edge of the science of their era. Also, your astrological star sign has been handed down to you from ancient times. What does it mean? Read on to find out.

About the Celestial Sphere

On a clear night with little or no moonlight, the stars in the sky are a breathtaking sight, particularly when seen from a location where the entire horizon is visible. If you live in Britain, take advantage of a clear night to view the stars as most nights the sky is often too cloudy to see the stars. When the night sky is clear and the entire horizon can be seen, the night sky is like a vast dark dome to which the stars are attached. It is not surprising that astronomers long ago imagined that all the stars were attached to an invisible sphere, surrounding the Earth, which they called the Celestial Sphere, as illustrated in Figure 1.1.

To an observer on the ground, the stars move across the sky at a rate of 15 degrees every hour. This is because the Earth spins through 360 degrees about an axis through its poles at a constant rate of once every 24 hours, corresponding to 15 degrees per hour. Ancient astronomers thought the Earth was fixed at the centre of the Celestial Sphere. Spinning at a constant rate of once every 24 hours, it carried the stars across the sky at a steady rate.

The Pole Star, known also as Polaris, is found directly above the Earth's North Pole. The axis of rotation of the Celestial Sphere can be imagined to pass through Polaris.

The Celestial Equator is the projection of the Earth's equator onto the Celestial Sphere. The stars in the night sky are in the opposite direction to the Sun. The Earth takes one year to orbit the Sun. (Anyone who spends a whole year at the Earth's equator would see the entire Celestial Sphere over the course of a year.)

ACTIVITY

1 Find Ursa Major from its pattern on page 7 and locate the Pole Star. Make a sketch of the Pole Star, Ursa Major and the horizon. Repeat the observation three hours later and add the new position of Ursa Major to your sketch. You should find that Ursa Major turns through 45 degrees in 3 hours.

2 You will need a camera with a shutter you can set to B for a long exposure for this activity. Point the camera towards the Pole Star and open the shutter for 20 minutes. When the film is developed, you should find concentric circular arcs centred on the Pole Star, as in Figure 3.1. Each arc has been traced out by the image of a star on the film because the camera moves relative to the star's position as the Earth turns.

Figure 3.1   Star trails (John Sanford/Science Photo Library)

Circumpolar stars

The Pole Star can be seen on any clear night in the Northern hemisphere at any time of the year. Stars near the Pole Star can also be seen on any clear night. The Earth's spinning motion makes each of these stars appear to move round on a circle centred on the Pole Star. The altitude of such a star changes as it moves round the Pole Star. If the star never dips below the horizon, the star is referred to as **circumpolar**. More and more stars are circumpolar the further North an observer is. (An observer at the Earth's North Pole can see every star North of the Equator on any clear night.)

Stars that rise and set

Stars that are not circumpolar rise and set once every 24 hours. For example, on a clear night in early winter in the Northern hemisphere, the Orion constellation is seen just after it has risen above the Eastern horizon. By early morning before sunrise, the same constellation is seen above the Western horizon just before it sets. All stars that are not circumpolar rise in the East and set in the West because the Earth spins Eastwards about an axis through its poles.

Constellations for each season

The Earth's axis always points towards the Pole Star. As a result, the Earth's North Pole is tilted towards the Sun in June and away from the Sun in December. This is why mid-summer in the Northern hemisphere is in June and mid-winter is in December. If you live in the Southern hemisphere, you experience mid-winter in June because the South Pole is tilted away from the Sun then.

Different constellations are visible in the night sky during the year. This is because when we view them from the night side of the Earth they are in the opposite direction to the Sun. As the Earth orbits the Sun once each year our view of the night sky changes as the opposite direction to the Sun changes. For example, the Orion constellation is a glorious winter constellation in the Northern hemisphere because it is in the opposite direction to the Sun in winter. There is no point in looking for Orion in summer as it is in the same direction as the Sun because the Earth has moved round its orbit by about 180 degrees from its winter position. You can work out which constellations can be seen in each season using a star chart.

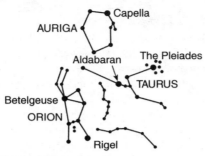

Figure 3.2   Part of a star chart

ACTIVITY

The next time the night sky is clear, use a star chart to identify which constellations ought to be visible. Then see if you can locate them in the night sky.

The ecliptic

If the Sun was much fainter, we would see the stars during the day. The Sun would move gradually through the constellations of the Zodiac over the course of a year. As explained on page 8, the Zodiac is a belt of constellations through which the planets move.

Imagine the Earth's orbit expanded onto the Celestial Sphere. This is the path the Sun would be seen moving along as viewed from the Earth if the Sun was much fainter. This path is called the **ecliptic**. The constellations of the Zodiac are the constellations along the ecliptic.

■ At mid-summer in the Northern hemisphere, the Sun reaches its highest point on the ecliptic north of the Celestial Equator. This is why the Sun is at its highest due South at midday at mid-summer. This time of the year is known as the **summer solstice** and it occurs when the Sun lies in the constellation of Cancer.

■ At mid-autumn in the Northern hemisphere, the Sun has moved round the ecliptic from its mid-summer position by 90 degrees. At this time of the year, it passes from north to south across the Celestial Equator in the constellation of Libra. This point of the year is known as the **autumn equinox**.

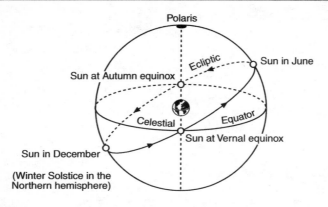

Figure 3.3   The ecliptic

■ At mid-winter in the Northern hemisphere, the Sun reaches its highest point on the ecliptic south of the Celestial Equator. At this time of the year, known as the **winter solstice**, the Sun lies in the constellation of Capricorn. At midday in mid-winter in the Northern hemisphere, the Sun is due South at its lowest point in the sky.

■ At mid-spring in the Northern hemisphere, the Sun passes across the Celestial Equator from south to north in the constellation of Aries. This point of the year is known as the **vernal equinox**.

What's in a name?

Equinox means equal day and night. At mid-spring and mid-autumn, darkness and daylight each day are of equal duration.

Solstice means the Sun is 'still'. At mid-summer and mid-winter, the Sun is furthest from the Celestial Equator and its midday height changes very slowly.

The precession of the equinoxes

The Earth spins at a steady rate about an axis that points to the Pole Star; its equator does not lie in the same plane as its orbit. Imagine a model of the Earth and the Sun in which the Earth's equator was in the same plane as its orbit. The Earth would be a very boring and predictable place. Every day would consist of 12 hours of daylight and 12 hours of darkness. There would be no seasons either and so it would be difficult to keep track of the year. In our model, the Earth's axis would need to be tilted by 23.5 degrees to represent the real Earth on its orbit round the Sun. This is the actual angle between the ecliptic and the Celestial Equator.

The direction in which the Earth's axis points is gradually changing. By the next millennium in the year 3000, the Earth's axis will have turned through about 35 degrees. This effect, known as **precession**, can be seen when a child's spinning top is tilted, and its axis slowly turns about a vertical line through its pointed base. The Earth is precessing at a rate of once every 26 000 years. As a result, the bright star Vega and not Polaris will lie close to the direction the Earth's North Pole points in 13 000 years' time. The equinoxes and solstices are gradually moving along the ecliptic at a rate of about 1 degree every 70 years. In 13 000 years' time, the vernal equinox will be where the autumn equinox now is and vice versa.

You and your star sign

What is your star sign? Do you know its astronomical meaning? Sunrise is like the birth of a new day. The constellation the Sun lies in on your birthday is your star sign – or at least it would have been if you had lived about 3000 BC when the constellations were first linked to the seasons of the year. For example, the Sun lay in Aries in mid-spring in 3000 BC. Now its mid-spring position is in Pisces due to the gradual precession of the Earth's axis. Anyone with a birthday in mid-March will find his or her fortunes predicted under Aries even though the Sun is in Pisces in mid-March now. Star signs and birth dates provide entertainment for many people but there is no scientific evidence that humans are affected physically by the stars. Whether or not individuals can unknowingly adopt a personality corresponding to their star sign is another matter. For example, a person born under Libra might see himself or herself as 'well balanced' because that is a supposedly Libran characteristic. The fact that many people still take astrology seriously clearly indicates just how deep-rooted the subject was centuries ago.

Star coordinates

The position of a star in the night sky changes by 15 degrees every hour because of the Earth's rotation. A star that is due South high in the sky at midnight gradually moves westwards after midnight. Its elevation above

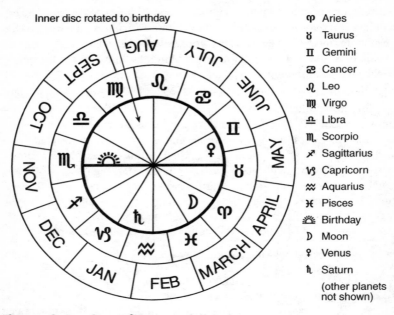

♈	Aries
♉	Taurus
♊	Gemini
♋	Cancer
♌	Leo
♍	Virgo
♎	Libra
♏	Scorpio
♐	Sagittarius
♑	Capricorn
♒	Aquarius
♓	Pisces
☼	Birthday
☽	Moon
♀	Venus
♄	Saturn
	(other planets not shown)

Figure 3.4   Star signs

the horizon also changes. Because its position relative to all the other stars does not change, its position is defined with reference to a point on the Celestial Equator called the **First Point of Aries**. This is where the Sun crosses the Celestial Equator as it moves from South to North along the ecliptic, its annual path through the constellations. See page 20.

Every star lies on a great circle through the poles. Two coordinates are needed to locate a star on the Celestial Sphere, as shown in Figure 3.5.

■ **The declination of a star** is like the latitude of a point on the Earth (which is defined as the angle between the equator and that point to the centre of the Earth). The declination of a star is the angle of a star, north or south of the Celestial Equator.

■ **The right ascension of a star** is the angle along the Celestial Equator from the First Point of Aries to the great circle through the star. Because the Earth spins once every 24 hours, each star reaches its highest point in the sky (or 'culminates' to use the technical term) once every 24 hours. Right ascension may be expressed in hours and minutes as a star moves through 15 degrees every hour. The right ascension of a star, in hours and minutes, is therefore the time from the culmination of the First Point of Aries to the culmination of the star. For example, a star with a right ascension of 10 hours and 30 minutes culminates 10 hours and 30 minutes after the First Point of Aries culminates.

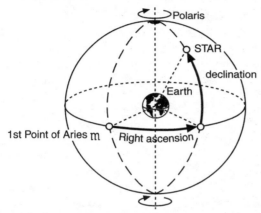

Figure 3.5   Star coordinates

ACTIVITY

Use a marker pen to draw a circle round a ball to divide its surface into two hemispheres. If the circle represents the Celestial Equator, mark the position of the Pole Star on the ball and any position along the circle to represent the First Point of Aries. Mark a star on the ball and draw the great circle through it and both poles. Make the northern hemisphere of the ball turn clockwise once every 24 seconds, to represent the spinning motion of the Celestial Sphere. Each second represents one hour of a full day. View the ball at eye level as you make it turn. The time between the First Point of Aries and the great circle through the marked star passing in front of you represents the right ascension of the star.

The first cosmologists

Prominent constellations such as Ursa Major and Orion were first recorded over 4500 years ago. In 125 BC, an astronomer in Greece called Hipparchus analysed an earlier description of the constellations. He compared the description with his own observations and concluded that the equinoxes had moved gradually. This impressive scientific analysis demonstrates the power of human thought and imagination that existed long before the scientific age in which we live. Hipparchus is also credited with adopting the Babylonian method of dividing a circle into 360 degrees.

Hipparchus was undoubtedly influenced by Aristotle (384–322 BC) who established his authority on science so firmly that his methods lasted thousands of years. Aristotle was a pupil of Plato, the famous philosopher, at the Academy in Athens. Plato's idealism perhaps explains why observations played little part in Aristotle's scientific models. In his cosmological work *On the heavens* he wrote that 'Heaven must be a sphere, for this is the only form worthy of its essence...' He considered the Earth and earthly things as imperfect unlike the divine nature of Heaven. He placed the stars and planets in circular orbits about the Earth. Aristotle's ideas on cosmology influenced religion for many centuries and those who disagreed with his teachings were considered to be heretics. Aristotle's approach to physics and astronomy centred on a search for coherence with his philosophical ideas. He imagined matter to be composed of the four so-called elements, earth, water, air and fire. The stars and planets were thought to be composed of a fifth element which was present only in the heavens.

Aristotle's ideas were challenged in about 300 BC by Aristarchus who put forward the idea that the Earth and planets orbit the Sun. His imaginative model was rejected on several grounds including the observation that the positions of the stars did not seem to be affected by the Earth's position. Aristarchus' opponents argued that if the Earth moves round the Sun, the stars should change their positions relative to each other when observed from different positions on the Earth's orbit. This effect, known as **parallax**, does occur but was too small to be detected by astronomers until the eighteenth century. It was also reasoned that a spinning Earth should throw objects off its surface. In addition, the Earth would have been removed from its special place in the Universe if it was in orbit round the Sun.

Ptolemy's model of the Universe

Aristotle's cosmological model was developed further by Ptolemy who lived in Egypt in the second century AD. Ptolemy wrote a great work or *Almagest* consisting of 11 books which covered everything known about astronomy. His catalogue of 1022 stars was not surpassed for three centuries. He is best remembered by astronomers for his model of the Universe which placed the Earth at the centre with each planet rotating in an **epicycle**, a circle whose centre circles the Earth. Figure 3.6 shows the idea.

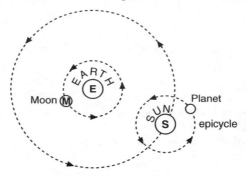

Figure 3.6   Ptolemy's model

Ptolemy devised this model of the Universe to explain the **retrograde motion** of the outer planets. This motion is evident if Mars or Jupiter's path through the constellations of the night sky is plotted for several weeks before and after opposition (when it is in the opposite direction to the Sun). The planet's nightly progress from west to east through the constellations reverses for a few weeks then resumes from west to east. This reversal is referred to as retrograde motion.

The reason why retrograde motion occurs is because the Earth itself is a planet in orbit round the Sun. An outer planet is further from the Sun than the Earth so it moves more slowly than the Earth. Each time the Earth catches up and overtakes an outer planet, the outer planet's position against the distant stars reverses then resumes its original direction of motion. Ptolemy believed the Earth to be fixed so his observation of retrograde motion presented him with the problem of how to explain it in terms of a fixed Earth. His 'epicycle' solution stood for centuries, in support of the Church's view that the Earth and its inhabitants occupy a privileged position in the Universe at its centre.

Ptolemy ignored the heliocentric model of Aristarchus. It was forgotten until rediscovered almost fifteen centuries later in the sixteenth century by a Polish monk called Copernicus. However, Copernicus did not publicize his researches, perhaps fearful of the wrath of the Church. In 1632 Galileo, the most famous scientist of the age, published the case for the Copernican model, having gathered sufficient observational evidence to realize that Copernicus was right. The Earth does move round the Sun and has no special place in the Universe. However, just as in our modern scientific age, cosmological questions in ancient times were of profound significance and were at the very core of the philosophy of the age.

Summary

- **Circumpolar stars** never set and can be seen at any time of year on a clear night.
- **Constellations that rise and set** change during the year as the Earth moves round the Sun.
- **The ecliptic** is the apparent path of the Sun through the constellations. It is the Earth's orbit projected on the Celestial Sphere.
- **The equinoxes** are at mid-spring and mid-autumn, corresponding to the two points of intersection of the ecliptic and the Celestial Equator. The First Point of Aries is the point of intersection corresponding to the vernal or spring equinox.
- **The declination** of a star is the angle between the star and the Celestial Equator.
- **The right ascension** of a star is the angle along the Celestial Equator from the great circle through the star to the First Point of Aries.
- **Opposition** is when a star or planet is in the opposite direction to the Sun. It's the best time to see it.
- **Retrograde motion** occurs when the progress of an outer planet through the constellations reverses for a period of weeks before resuming its normal progress.

4 THE SCIENTIFIC REVOLUTION

The Scientific Age in which we live began several centuries ago. Perhaps the discovery of the New World made scientists and philosophers think about nature differently to those who accepted Aristotle's teachings. Galileo may be considered to be the father of modern science because he demonstrated that Aristotle's way of doing science by ignoring the crucial role of observations was fundamentally flawed. Over succeeding centuries, Galileo's approach to science was developed and the link between laws and observations became evident as new discoveries were made, analysed and tested. The modern philosophy of science is that laws and theories hold on the basis of never having being disproved. This philosophy was formulated by Sir Karl Popper in the twentieth century who realized that observations and measurements can never prove a theory or a law but at any time, a single experiment is sufficient to disprove a theory or law.

Scientific controversies

Before the Scientific Age, many theories of science were based on the assumption that the Earth is at the centre of the Universe and that living beings were created by one or more superior beings who designated a special role for humans. The idea that humans evolved from apes over thousands of centuries found little favour, as it omitted the role of a creator. Theories about the natural world were usually chosen on grounds we would consider unscientific and selected facts were used to support the theories. Other facts that did not match the theories were discarded as unreliable or imperfect. Not surprisingly, alchemy and astrology were two major strands of scientific endeavour before Galileo. For example, attempts to turn lead into gold or to predict events occupied the working lives of many individuals, undoubtedly financed by rich and powerful patrons who imagined that they would amass further wealth as a result of such activities. Aristotle's approach of picking facts to support theory dominated the way science was conducted long after Aristotle's death, from Ancient Greece, through the Roman Empire, the Dark Ages and into the Middle Ages.

Copernicus, an unlikely revolutionary

The search for simplicity in nature is perhaps understandable when we look around and see just how varied the natural world is. One of the reasons why the Earth was considered to be at the centre of the Universe is that objects fall to the ground when released. Because a solid object is more dense than water or air, it was supposed that all solid material was in or on the Earth at the centre of the Universe. The spherical shape of the Earth had been deduced from astronomical observations such as the fact that the altitude of the Pole Star increases the further North the observer is. The rotation of the Celestial Sphere was deduced from the daily rising and setting of constellations. The other possibility that the Earth spins and the Celestial Sphere did not was discounted on the grounds that such a rapid spinning motion would cause air to rush past heavy objects on the equator and cause lighter objects to be thrown off. Aristotle's theory that the stars are spinning round the Earth neglected to consider why the stars themselves were not affected by their spinning motion, especially since such motion would be even faster than on the Earth due to the much greater distances to be covered each day.

Copernicus was a Polish-born medieval scholar whose interests included astronomy, economics, mathematics and medicine. He devoted many years of his life to studying ancient works to find out why objections to Ptolemy's model of the Universe had been rejected. He constructed his own version of Ptolemy's model and found it was necessary to have more than thirty spheres to account for all the observations of the motion of the stars and the planets. He realized such complexity was not necessary if the Sun and not the Earth was at the centre, and he reduced the model to little more than a set of concentric circles representing the orbits of the planets. Copernicus feared that his revolutionary ideas would be ridiculed by his contemporaries and his major work *De revolutionibus orbium coelestium* was not published until shortly before he died in 1543. It was presented as a mathematical challenge rather than a new scientific theory and it failed to have any immediate impact. However, in 1600, Giordano Bruno was burned at the stake by the established Church for promoting the

Copernican model and his own view that the Universe is not finite, as the Celestial Sphere model implies. Surprisingly, the Copernican model was not fully accepted by the Church until the early nineteenth century. In contrast, the Church now is far more willing to listen and has organized scientific conferences on current developments in cosmology.

One of Copernicus' opponents was Tycho Brahe who developed a world-famous reputation for practical astronomy. At his observatory in Denmark under royal patronage, Brahe and his assistants catalogued over 700 stars (before the invention of the telescope) using instruments he devised that were capable of measuring star positions to an accuracy of less than a sixtieth of a degree. Such accuracy far exceeded previous measurements and was not bettered for many years after his death. Even with such accurate instruments, the change of position of a star due to the Earth's orbital motion proved too small. He strongly supported Ptolemy's model, perhaps rejecting the new ideas put forward by Copernicus as he was unable to obtain direct evidence for the motion of the Earth through space. Nevertheless, his observation of a new star in 1572 did challenge Aristotle's idea that the stars are forever fixed to a celestial sphere. From careful observations on this new star, he concluded that this star had never been seen before in any age since the world began. Tycho Brahe moved to Prague in 1597, four years before he died, where he took on Johann Kepler as his assistant. Kepler was a Copernican and he examined Brahe's record of observations very carefully as well as making his own detailed observations. He catalogued over 200 more stars and he deduced a series of three laws that describe the motion of the planets. These laws were not successfully explained until many decades later when Sir Isaac Newton devised his theory of gravity.

Kepler, the law maker

In his early research, Kepler measured the distance from each planet to the Sun and he discovered that the distance from the Sun to each planet changes. In the Copernican model, each planet moved round the Sun on a circular orbit, the Sun being at the centre of the circle. For a circular orbit, the distance from the Sun to the planet is therefore constant. Kepler modified the Copernican model, initially by describing three circles for each planet, an outer circle, a median circle and an inner circle. He used his knowledge of geometry to show that orbits of the planets including the Earth can be linked to each other by means of regular shapes. For example, if the orbit of Saturn was on the surface of a sphere, the orbit of Jupiter would just touch the surfaces of a cube that just fits into the sphere. Likewise, if the orbit of Jupiter was on a different sphere, the orbit of Mars would just touch the surfaces of a tetrahedron that fits into the sphere. Kepler discovered other shapes to fit the orbits of Earth, Venus and Mercury. However, he was unable to obtain a consistent fit as he found that the orbits were a little too large in some places and a little too small in other places.

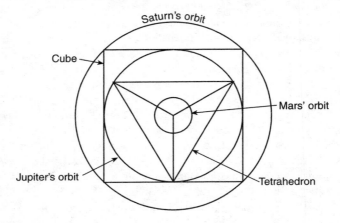

Figure 4.1   Kepler's geometrical models

Perhaps Kepler realized that the role of geometry and shapes could not be developed much further at this stage even though the appeal of geometry to astronomers had a very long history. He returned to study the detailed observations on the path of Mars made by Tycho Brahe and realized that the Copernican model needed to be modified to explain the path of Mars. He realized that the orbit of Mars is an ellipse, with the Sun at one of the two 'focal points' of the ellipse, as explained in Figure 4.2. He went on to establish the general rule, known as Kepler's 1st Law, that all the planets move on elliptical orbits of differing ellipticity. He found that the idea of three circles for each planet corresponded to the closest, furthest and mean distance of the planet from the Sun on its elliptical orbit. He also established two further laws which describe how fast planets move round their orbits and how long each planet takes to orbit the Sun once. We will look in detail at these important laws in the next section of this chapter.

ACTIVITIES

1 Make a geometrical model of the orbits of Saturn and Jupiter

The orbits of Saturn and of Jupiter are about nine times and five times bigger than the Earth's orbit.

Draw a circle with a square inside it such that the corners of the square all touch the circle. To do this, draw two perpendicular diameters to give the corners of the square and then complete the square.

Then draw another circle inside the square so it touches each side of the square. The ratio of the diameters of the two circles should be about the same as the ratio of the orbits of Saturn and Jupiter.

2 Draw an ellipse

You will need two thumb-tacks, a tablemat (to stop the thumb-tacks penetrating the table surface), string and a blank sheet of paper. Place the mat under the paper and push the thumb-tacks into the mat through the paper. The thumb-tacks should be about 5 cm apart. Make a loop of string of total length about 15 cm. Place the loop round the thumb-tacks and pull it taut on the sheet using the sharp end of the pencil. Keeping the loop taut, draw an ellipse by moving the pencil round the loop, as in Figure 4.2. The thumb-tacks are at the two focal points of the ellipse. To represent a planet's orbit, the Sun needs to be represented at one of these two points. The distance to the planet therefore depends on its position on its orbit.

■ The perihelion is the least distance of the planet to the Sun.
■ The aphelion is the greatest distance of the planet to the Sun.

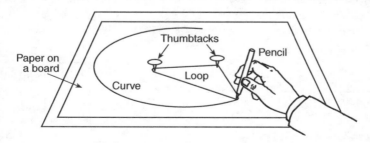

Figure 4.2   Making an ellipse

Kepler's laws of planetary motion

In addition to measuring the orbit of each planet, Kepler measured the progress of each planet as it went round the Sun and also determined how long each planet took to make one complete orbit of the Sun. He expressed the distance measurements in astronomical units (1 AU is the mean distance from the Sun to the Earth) and the times in days and years. From his measurements, he formulated three laws to describe the motion of the planets. As we have seen, the first law is essentially geometrical, and states that each planet moves along an elliptical orbit in which the Sun is at one of the two focal points of the ellipse.

However, the other two laws express numerical relationships and they represent the first significant attempt to describe stars and planets in mathematical terms rather than geometrically.

The second law established by Kepler was deduced from the accurate measurements of Mars made by Tycho Brahe. Kepler discovered that the speed of Mars at perihelion was greater than its speed at aphelion. More precisely, he determined the angle the line from the Sun to Mars moved through over the same period of time at perihelion and compared this with the same measure at aphelion. He knew that the perihelion distance of Mars is 0.9 × the aphelion distance. He found the planet's apparent progress at aphelion was 0.81 × its apparent progress at perihelion which he realized is the same as the square of the perihelion distance to the aphelion distance. More generally, Kepler deduced that the rate of progress of an imaginary line from the Sun to a planet varies as the inverse of the square of the distance from the planet to the Sun.

Kepler then turned his attention to comparing the orbits of the planets. He knew the mean distance of each planet from the Sun in terms of the mean distance from the Sun to the Earth and he knew how long each planet took to make one complete orbit (i.e. its period of orbit). The data is given in the table below.

	Mercury	Venus	Earth	Mars	Jupiter	Saturn
Period of orbit, P (in years)	0.2(4)	0.6(1)	1.0	1.9	11.9	29.5
Mean radius of orbit, R (in AU)	0.3(9)	0.7(2)	1.0	1.5	5.2	9.5

What can you deduce from this data? Clearly, the greater the mean distance of a planet from the Sun, the longer it takes to go round. Can you deduce a numerical link as Kepler did? A graph to show the relationship between P and R is shown in Figure 4.3. This shows the relationship is non-linear as the period is not proportional to the radius. In other words, Jupiter is five times further from the Sun than the Earth but it takes almost twelve times as long as the Earth to go round the Sun once.

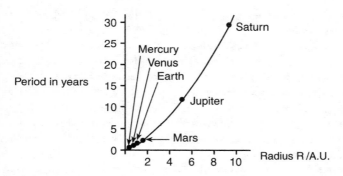

Figure 4.3   Period v mean radius

Kepler realized that the cube of the mean radius is the same as the square of the period for each planet. In simple mathematical terms, using the units in the table, this is usually written as an equation, known as Kepler's 3rd Law.

$$(\text{period})^2 = (\text{mean radius})^3$$

The relationship has been worked out for two of the planets in the table below. Prove for yourself that the law holds for the other planets too.

	Mercury	Venus	Earth	Mars	Jupiter	Saturn
Mean radius R/AU	0.3(9)	0.7(2)	1.0	1.5	5.2	9.5
Period P/years	0.2(4)	0.6(1)	1.0	1.9	11.9	29.5
R³		0.3(7)			141.0	
P²		0.3(7)			141.0	

Kepler was unable to explain his three laws of planetary motion. He demonstrated that the geometrical patterns that his predecessors sought to establish needed to be replaced by mathematical relationships in the form of his laws of planetary motion. He replaced the harmony of spheres and circles as a description of the motion of the planets, a philosophy that had underpinned the work of astronomers for many centuries, with an elegant mathematical description of the motion of the planets.

Galileo Galilei

More scientific discoveries have been made in the past hundred years than ever before. Science came of age in the twentieth century and will undoubtedly form the basis of human activity and culture in the future. One can only speculate on the discoveries that lie ahead as no one knows how much more there is to know! However, it is not unreasonable to predict that scientific discoveries will continue to be made and it is certain that some of these discoveries will affect the human race very significantly. Four centuries ago, such thoughts would probably have been dismissed as heresy and those who promoted such views could have been put on trial. The change from the authoritarian and rigid theories of the medieval age to the experimental philosophy that underpins our Scientific Age may not have originated with Galileo, but there is little doubt that Galileo showed, through his discoveries, the power of science to change established views. Equally important, he clearly demonstrated the significance of reliable and accurate observations. However, perhaps his most important contribution was to establish that science has limits which cannot be extended without valid evidence. In showing that experimental science should be used to explain what we observe, he cut the link between physical science and religion that had existed for centuries. The link continued in biological science well into the nineteenth century through the controversies surrounding Charles Darwin's theory of evolution. Even in this field, the modern science of genetics which supports evolution has grown enormously through scientific methods developed from Galileo's approach to science.

Galileo was born in 1564 in Pisa, Italy, the same year in which William Shakespeare was born. As the son of a nobleman, Galileo was educated in a monastery. In 1595 he became Professor of Mathematics at the University of Padua, one of Europe's leading universities at that time, in what was then the Republic of Venice. His Venetian paymasters allowed him to follow his own interests and his discoveries on motion would have been sufficient to win long-lasting recognition. From time to time, he made scientific instruments for commercial purposes such as measuring the density of precious metals and stones. In 1609, reports reached him about the invention of an optical device, the telescope, for making distant objects appear larger and closer. Within a short time, he had designed and constructed his own telescope, capable of making distant objects appear ten times larger and closer. He demonstrated the power of his telescope to the Senate at the top of the Campanile in Venice by observing incoming ships fifty miles away, two or more days' sailing away from the port. Such information was very valuable to the merchants and brokers of Venice.

Galileo realized he could use the telescope to study heavenly bodies. He wanted to gather evidence in support of the Copernican model and he devised a more powerful telescope, capable of magnifying objects thirty times. He was astounded when he first used this telescope to study the night sky. He observed ten times as many stars as can be seen directly without a telescope. He found that the surface of the Moon is heavily cratered and he discovered the four innermost moons of Jupiter which he called the Medicean planets in honour of the Medici family, one of his wealthy sponsors. Nowadays, these four moons are more usually called the galilean satellites of Jupiter.

ACTIVITY

Following Galileo's footsteps

Stars too faint to be seen with the naked eye can be seen using a telescope or binoculars. This is because a telescope is much wider than the pupil of the eye so it collects much more light than the eye does. Observe a constellation with and without binoculars or a telescope. In each case, count the number of visible stars and you should find many more can be seen when you use a telescope or binoculars.

The eye pupil at its widest is no more than about 10 mm in diameter. A telescope of diameter 100 mm has an area of cross-section 100 times greater than the area of a 10 mm diameter eye pupil. Prove this for yourself using the formula 'area of a circle = $\frac{1}{4} \pi \times$ diameter^2'. The amount of light collected is proportional to the entrance area of the telescope or the eye. Thus a telescope of diameter 100 mm collects 100 times more light than an eye pupil of diameter 10 mm.

The lunar surface was found by Galileo to be much rougher than anyone imagined. The light and dark patches visible without the aid of a telescope were covered in spots, particularly the lighter areas. As the Moon passes through its regular cycle of phases, the boundary between sunlight and shadow moves across the surface. Galileo observed that the boundary itself was slightly irregular and saw that many of the spots at the boundary were flooded with light on the side away from the Sun. Galileo realized that the irregularities in the boundary were there because the lunar surface is uneven. The uneven illumination of the spots near the boundary occurs because these spots are hollows or craters or small mountains. Viewed from the Earth, the inside of a crater at the boundary would be dark on the side nearest the Sun and bright on the side facing the Sun.

> Use binoculars or a telescope to observe the boundary between the bright part and the dark part of the lunar surface and see if you can repeat Galileo's observations.

The four innermost moons of Jupiter were discovered by Galileo as a result of telescopic observations in early 1610. One night in January, he noticed three faint points of light near Jupiter which he had not seen before. He was struck by the fact that they were in a straight line parallel to the ecliptic, the path of the Sun across the sky. Two of these points were east of Jupiter and the other was on its western side. He repeated his observations the next night and found the three points were now all west of Jupiter. He realized that the three points were probably moons in orbit round Jupiter but he had to wait for two days before clear skies enabled him to make further observations. The next time he observed Jupiter, he could only see two of the three moons, but still both in the same line as before, and he guessed that the third moon was hidden behind Jupiter itself. Further observations on the brightness and positions of the moons led Galileo to conclude that Jupiter has four moons. He named them Io, Callisto, Ganymede and Europa.

> Use binoculars or a telescope to see if you can observe Jupiter's moons. You will need to wait until Jupiter is approximately in the opposite direction from the Sun which occurs for a few months every year.

Galileo on trial

Galileo's astronomical discoveries were widely reported in Europe. His Copernican views became known to the Church who opened a file on him. Galileo hoped his observations and conclusions in support of the Copernican model would be accepted by the Church so he would no longer need to rely on the protection of the anti-clerical Venetian authorities. In 1613, he was reprimanded by the Church for his views. Three years later, aware of the continuing disapproval of the Church, he travelled to Rome to attempt to persuade sympathetic cardinals to support the Copernican system. He was to be disappointed and he returned to Florence, which was now his home, convinced that his ideas would eventually prevail. In 1623, a new pope was elected who had a reputation as an intellectual. Once again, Galileo travelled to Rome to try to persuade the new pope, Urban VIII, and his officials to withdraw the ban on the Copernican system imposed in 1616. Again, Galileo was to be disappointed, for the new pope refused to accept any challenge to the existing order. Galileo returned once more to Florence to search for a different way forward. He decided to set out his views and his support for the Copernican model in print in Italian, for all to read. He had completed his work, *Dialogue on the Two Chief World Systems*, by 1630 and after some difficulties with the censor authorities, it was published in Florence in 1632. He presented his work as a dialogue between two main characters, Salviata and Sagredo. Salviata was presented as a Copernican whose ideas were questioned by Sagredo as an intelligent non-scientist. The Ptolemaic system was presented by a third character, Simplicius. The sample of the dialogue presented below shows how Galileo managed to convince the censors that he was not presenting a challenge to the Church.

Sagredo   *What is the reason for calling the four Jovian planets 'moons'?*

Salviata   *That is how they would appear to anyone who saw them from Jupiter. They always look entirely illuminated to us who are closer to the Sun. Anyone on Jupiter would see phases of these moons just like we see phases of our own moon.*

Simplicius   *Ptolemy must have known about this and found a way in his system to explain it.*

Galileo's book was an instant best-seller and the Church reacted rapidly by banning it and stopping further reprints. Galileo, now in his seventieth year, was summoned to appear before the Holy Office of the Inquisition in Rome on the 12th of April, 1633. Galileo duly appeared in front of the twelve judges of the Inquisition. No consideration was given to the discoveries made by Galileo. Instead, the judges sought to show that Galileo had breached the agreement of 1616 when he was instructed not to hold or defend the Copernican view. His defence was that the ban imposed in 1616 did not prevent teaching the Copernican view. He argued that his dialogue presented both views to the reader. The judges decided that Galileo had broken the 1616 ban and had acted deceitfully. He was threatened with torture, forced to recant and made to spend the rest of his life under house arrest at his home in Florence where he died in 1642. Science as a high-profile activity came to a standstill in Catholic Europe for many years. However, Galileo's dialogue and his discoveries were taken up vigorously by scientists in Northern Europe where the Church had much less authority. The publicity that Galileo's trial attracted perhaps helped to promote the Copernican system. Galileo's immense contribution to the development of science cannot be understated.

Newton's Universe

Galileo's ideas, Kepler's laws and the Copernican system were brought together by Sir Isaac Newton who developed laws of motion that underpin much of modern technology and engineering. Newton was born in 1642, the same year that Galileo died, in the market town of Grantham which is in the county of Lincolnshire in eastern England. His father died before he was born, and he was brought up by a grandparent after his mother remarried. He was sent to the local grammar school as a boarder and entered Cambridge University in 1661. England at this time was a republic under Oliver Cromwell. The University was closed at times during 1665 and 1666 because of the Great Plague which ravaged the country during those years. Newton returned to his home in the Lincolnshire countryside and in just two years produced mathematical theorems and physical theories including his law of gravitation that revolutionized mathematics and physics. He returned to Cambridge in 1667 and was appointed two years later at 26 years of age to the Chair of Mathematics at Trinity College.

Newton's interests in science were wide-ranging, and included astronomy, chemistry and optics as well as mathematics and physics. He set out his theories of mathematics and physics in his greatest work, the Principia, in which he showed that his three laws of motion and his law of gravitation are sufficient to explain the motion of bodies. He proved once and for all that the planets and the Earth orbit the Sun and he explained Kepler's laws and Galileo's laws. Using his law of gravitation, he was able to predict comets, eclipses and tides. He saw the Universe as a gigantic mechanical system, governed by the same laws that govern the motion of objects on the Earth. His ideas provided the guiding principles for science for the next two centuries until Einstein showed that space and time are not independent quantities. We will meet Einstein's ideas in detail in Chapter 9. Newton published the Principia in 1687 and he established himself as the foremost scientist of his generation. Sadly, he became involved in a bitter dispute with Leibnitz who claimed to have invented calculus before Newton. At the University, promotion was blocked because he belonged to the Unitarian Church and did not believe in the Holy Trinity. He was therefore not eligible to become a cleric in the Church of England, an essential element of the 'person specification' for a University appointment in the seventeenth century. He left the University in 1696 to become Master of the Mint where he devoted his talents to monetary reform, only delving back to the challenges of science occasionally. His pre-eminence as a scientist was recognized in 1703 when he was elected as President of the Royal Society and he was knighted in 1705. In contrast to Galileo, who was shunned by the Catholic Church, Newton became part of the Establishment in England, even attracting the attention of satirists, clearly unaware that the freedom of thought on which they depended would not have emerged without Galileo and Newton.

ACTIVITY

Here are some simple experiments to demonstrate Newton's three laws of motion.

Newton's 1st law of motion An object continues at rest or in uniform motion unless acted on by a force.

Flick a coin across a very smooth level surface (e.g. a tea tray) and it will slide across the surface. In the absence of friction, the coin continues to move without being pushed because no force acts on it.

Newton's 2nd law of motion The force on an object is proportional to its rate of change of momentum.

Release a coin at the top of a flat slope (e.g. a tea tray propped up at one end) so that it slides down the slope. Observe it gather speed as it moves down the slope. Its momentum increases because its weight provides a steady force that pulls it down the slope.

Newton's 3rd law of motion When two bodies interact, they exert equal and opposite forces on each other.

Clap your hands together. The force of your left hand on your right hand is equal and opposite to the force of your right hand on your left hand.

Newton's law of gravitation

Before Newton produced his universal law of gravitation, it was generally believed that objects possessed gravity which pulled down and levity which pushed up. A falling object had more gravity than levity. Even now, the word 'levitation' is sometimes used for an object when it floats. Newton showed that a force of gravitational attraction exists between any two objects. He explained the motion of an object falling to the ground by saying that the object and the Earth attract each other. He used the same idea to explain why the Moon goes round the Earth and why the planets go round the Sun. If the force of gravity between the Sun and the planets suddenly ceased to exist, each planet would continue in uniform motion in a straight line at a tangent to its orbit. The force of gravitational attraction between the planet and the Sun keeps the planet circling the Sun. Newton showed that the force of gravity between two point objects was proportional to

■ the mass of each object
■ the inverse of the square of the distance between the two objects.

For two point objects of masses m_1 and m_2 at distance apart r, he formed the following equation for the force of gravity F between the two masses.

$$F = G \frac{m_1 m_2}{r^2}$$

where G is a constant which he called the Universal Constant of Gravitation.

Newton's choice of r^2 in his equation rather than r or r^3 or some other power of r was inspired by his previous discovery of the laws of motion. He knew that an object accelerated when it was acted on by a force and he had worked out that a body in steady circular motion always experienced an acceleration towards the centre of the circle. He had worked out that this acceleration was equal to the $(speed)^2$/radius. By linking this to his force formula, he derived Kepler's 3rd Law of planetary motion which we met on page 33. Any other power of r in his force formula would not have given Kepler's 3rd Law. A simplified version of Newton's proof of Kepler's 3rd Law is set out in Appendix 1 at the end of the book.

Newton's next step was to try to extend his ideas beyond point objects. This turned out to be very difficult and eventually, after many years, he proved that his law of gravitation could be applied to any two objects provided the distance in the equation was the distance between their centres of gravity. What did Newton have to say about the Universe at large? We will return to this question in the next chapter to see why Newton thought the Universe to be infinite in space and in time.

Laws and theories

Newton's laws of motion and his law of gravitation provided the theoretical foundations of physics and astronomy for over two centuries. Predictions based on Newton's laws proved to be accurate, leading to the discovery of two further planets beyond Uranus, which itself was not discovered until 1781, long after Newton's death. Following the discovery of Uranus by William Herschel, its path was plotted and calculations using Kepler's laws were made about its progress through the constellations. However, measurements on its progress showed that it did not follow Kepler's law exactly and its motion seemed to be affected by an undiscovered outer planet which was named Neptune. The position of this unknown planet was calculated by Urbain Le Verrier, a French mathematician, who used Newton's laws to pinpoint where astronomers should be looking to find Neptune. Le Verrier's predictions were justified when Neptune was indeed observed for the first time in 1846. Precise measurements of Neptune's position led Percival Lowell to deduce that another undiscovered planet was affecting Neptune and he predicted where this planet, Pluto, ought to be. Fourteen years after Lowell's death, Pluto was discovered by Clyde Tombaugh at the Lowell Observatory in Arizona. Both Neptune and Pluto were discovered as a result of predictions made using Newton's laws. Countless other predictions made using Newton's laws proved to be correct.

The laws set out by Newton seemed to provide a complete set of rules governing the motion of everything in the Universe. The scientific method established by Galileo was seen to be the correct way to pursue science. Let's consider the major steps in the scientific method:

1 An investigation often starts with a **hypothesis** which is essentially an unproven statement or prediction. For example, Lowell thought that there is an undiscovered planet beyond Neptune.

2 The next stage is to decide what **observations and measurements** need to be made. For example, long-exposure photographs of a limited area of the night sky were taken over a long period to locate Pluto.

3 Patterns and relationships deduced from the observations are used to draw **conclusions**. Tombaugh compared the stars on his photographs and discovered that one of the stars had changed its position relative to the others. Tombaugh concluded that this star could be a comet or a planet.

4 The conclusions are used to make **further predictions** which are then tested by making more observations and measurements, leading to further conclusions. If these conclusions continue to support the original hypothesis, the hypothesis gains credibility. For example, Tombaugh would have observed Pluto to see if it became brighter. If so, he would have concluded it was a comet approaching the Sun. Its brightness did not in fact change so he concluded it was indeed a planet.

The scientific method outlined above is used to establish scientific laws which are essentially generalized statements, often in mathematical form, about relationships between measurable quantities. Laws emerge from observations and measurements, often starting out as hypothetical relationships to be tested by experiments. Such a relationship assumes greater status if it is tested successfully. Further predictions leading to further tests boost its status more if the tests continue to support it. The hypothesis may eventually become accepted as a scientific law, in the form of either valid scientific statement or an accepted theory.

Experimental tests and observations have been used to formulate and establish scientific laws ever since Galileo. However, scientific laws can not be 'proved' like theorems of geometry. This is because if just one test fails to support a law, this is enough to disprove the law. In other words, **the laws of science are statements that have not been disproved**. This simple statement underpins a philosophy developed by Karl Popper in the twentieth century which is now accepted as how science works. By the end of the nineteenth century, many scientists thought that the major laws of science were all known and that science was nearing completion. The known laws of science were sufficient to explain observations and measurements in almost all branches of science. The notable exceptions were the results from experiments on heat radiation and the speed of light which did not fit the known laws of science.

Many scientists at the time thought these were minor problems which would be resolved in due course. However, some scientists working independently started to question some fundamental ideas which had been taken for granted in the laws of science. One of these scientists was Max Planck who devised the quantum theory to explain the spectrum of heat radiation from hot objects. The other scientist was Albert Einstein. Unlike Planck, who was an established University professor, Einstein was a newcomer when he wrote the Special Theory of Relativity, his first great work. Their theories revolutionized science within two or three decades. We shall look at their revolutionary ideas in more detail in the next chapter.

Summary

- ■ **The Copernican system:** the Sun not the Earth is at the centre of the system of planets. The planets including the Earth orbit the Sun.
- ■ **Kepler's laws of planetary motion**
  1 Each planet moves round the Sun on an elliptical orbit.
  2 The speed of each planet changes as it moves along its orbit. The progress of an imaginary line from the Sun to each planet varies as the inverse of the square of the distance from the planet to the Sun.
  3 The square of the period of a planet is proportional to the cube of its mean distance from the Sun.

■ **Galileo's astronomical discoveries** included
1 Observation of many stars too faint to see without the aid of a
   telescope
2 Observation of hollows, craters and small mountains on the Moon
3 Observation of the four innermost moons of Jupiter

■ **Newton's law of gravitation**
A force F of gravitational attraction exists between any two objects
and is proportional to:
1 the mass of each object
2 the inverse of the square of the distance between the centres of
   gravity of the two masses.

For two masses m_1 and m_2 spaced apart at distance r between their
centres of gravity

$$F = G \, \frac{m_1 \, m_2}{r^2}$$

where G is the Universal Constant of Gravitation.

5 | STRANGE IDEAS

The scientific laws and theories developed by Newton and his successors before the twentieth century made the world and beyond seem orderly and predictable. With these laws, the future state of any system of bodies could be predicted if its present state was known. In this chapter, we will look at the fundamental assumptions behind these laws. A grasp of these assumptions is essential to appreciate the successful challenges made to them by Einstein and Planck. The outcome of these challenges may seem strange, in defiance of everyday experience. For example, why should a fast-moving object appear shorter than if it was stationary? Why is energy at an atomic level 'lumpy' and not infinitely divisible? The theories put forward by Einstein and Planck have been successfully tested many times by many different experiments. In this chapter, we will look at key aspects of these theories that have influenced the work of astronomers and cosmologists, in preparation for later chapters where we shall consider the actual work in detail. Further fundamental questions have arisen from the theories of Einstein and Planck. Why is it not possible to travel faster than light, just as supersonic aircraft can travel faster than sound? Why should the mass of an object increase if it is supplied with energy? More knowledge about these matters will undoubtedly influence the development of cosmology. Bear this thought in mind as you read on!

Space and time

It's a well-known fact that the angles of a triangle add up to 180°. In fact, this is only true if the triangle is flat. To make a triangle, all you need to do is to draw straight lines between three dots on a piece of paper. If you make three dots on a ball and join the dots, you will finish up with a triangle which is not flat. In the nineteenth century, a German mathematician called Karl Gauss used surveying instruments to measure the angles of a very big triangle formed by three mountain tops. He wanted to see if the angles added up to 180° because he thought that space might not be flat on such a large scale. Perhaps he was disappointed to find the angles did add up to 180°. He had no evidence that space was not flat.

Newton did not question the rules of geometry that were first established thousands of years ago. He assumed that the shortest distance between any two points is a straight line. This would not be so if space is curved. Newton's laws assume the everyday rules of geometry and trigonometry which in turn assume that space is flat. Because Newton's laws successfully explain the motion of planets and comets, it is reasonable to assume that space is flat, at least on the scale of the Solar System. We shall see later that in fact Newton's laws do not completely explain the motion of the nearest planet to the Sun, Mercury. The Sun's gravity is strong enough to curve space just enough to affect Mercury's motion by a very tiny, but measurable amount. We shall return to this in Chapter 9 when we study black holes.

There is no evidence that Newton considered space as anything other than flat. He thought that space was flat, uniform and infinite, and absolutely the same here as anywhere else in the Universe. He also considered that time is absolute, running at the same rate everywhere in the Universe. Two timing devices synchronized to read the same time on Earth would, according to Newton, run at the same rate anywhere irrespective of their motion. If the two devices were brought together again after a long journey by one of them, they should read exactly the same. In fact, we shall see later that time is not an absolute quantity, and that a moving clock runs slower than a stationary one.

Can you tell if you are moving?

Most rail travellers have experienced the illusion of motion that occurs when their stationary train is next to an adjacent moving train. When the traveller's train begins to move again, for a moment, it's difficult to tell which train is moving – until the traveller feels a jolt. It would be impossible to tell if you were moving on a completely smooth train at night on a straight track when nothing can be seen outside. There would be no spills when you drank coffee and you could walk along the train without lurching about. Of course, real trains do not give completely smooth rides, and they speed up and slow down, so you could easily tell in practice if you were moving.

Newton's 1st law of motion states that an object continues at rest or in uniform motion if no forces act on it. If an object is being accelerated, there must be a force acting on it. For example, a falling object accelerates because it is attracted towards the Earth by gravity. We do not feel the Earth moving through space, so how can we detect if the Earth is moving? By measuring the position of the nearest stars at six-month intervals, it is possible to show that the Earth is in motion (we will look at this in detail in Chapter 6). The Earth moves round its orbit, a total distance of 942 million kilometres, in 365¼ days. Prove for yourself that it covers 2.580 million kilometres every day, which is more than 100 000 kilometres per hour or 30 kilometres every second, much faster than the fastest rocket or jet plane.

The question of trying to detect the Earth's motion through space without observing the stars was raised in the nineteenth century. The first accurate measurements of the speed of light had recently been made, and scientists knew that light travelled at a speed of 300 000 kilometres per second.

The experiment that didn't work

Two American scientists, Albert Michelson and Edward Morley, devised a special instrument to compare the speed of light moving parallel and perpendicular to the Earth's direction of motion. This instrument, called an interferometer, consisted of a beam splitter and two perpendicular arms of equal length, which each had a mirror at one end. Figure 5.1 shows a top view of this instrument. They knew that light consists of waves and they knew that if they were to split a beam of light into two perpendicular beams and then reunited the two beams, the waves of one beam would interfere with the waves of the other beam to produce a pattern of bright and dark fringes. Each dark fringe is formed as a result of the crests of the light waves of one beam being reunited with the troughs of the light waves of the other beam.

Michelson and Morley worked out that if the speed of light from the light source was affected by the Earth's motion, the fringe pattern would show a measurable shift when the apparatus was rotated through 90°. To see why, suppose the apparatus was positioned so one arm was parallel to the direction of the Earth's motion and then its length was adjusted so light would take the same time to travel along each arm and back. The two beams would reinforce to give a bright spot at the centre. Rotating the apparatus would reduce the time for one beam and increase it for the other

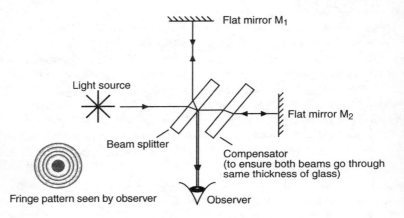

Figure 5.1   Michelson and Morley's experiment

so the bright spot would become dark then bright then dark etc as the two sets of waves moved in and out of phase with each other. Although the apparatus was easily capable of detecting the predicted shift, no shift was detected. The fact that the experiment gave a zero result caused great consternation amongst scientists. The failure to detect the effect of the Earth's motion on the speed of light shook the foundations of science that Newton and Galileo had established. Several ingenious explanations were proposed, including the idea that objects in motion contract physically along the direction of motion, but no evidence could be found to support these explanations.

ACTIVITY

Interference of waves

The principle of interference explains why a TV signal fades out when a metal tea tray is positioned near a portable TV aerial. Radio waves carrying the TV signal reach the aerial directly and by reflection off the metal tray. Fading happens when each crest of the radio waves reaches the receiver at the same time as a trough arriving via the other route. Try it.

What is light?

The nature of light is not easy to deduce. The many properties of light include reflection by mirrors, focusing by lenses and the formation of interference patterns as explained above. Most of its properties can be explained by assuming it is a waveform, travelling through space or through a transparent substance like waves travelling on the surface of water. In 1856, James Clerk Maxwell established the theory of electromagnetic waves in which he showed that electric and magnetic waves vibrating at right angles to each other would propagate through space at a speed of 300 000 km/s – the same speed as light. He concluded that light is therefore such a waveform which he referred to as an electromagnetic wave. He already knew that the wavelength of light waves depends on the colour of the light, ranging from about 0.4 micrometres for blue light to about 0.7 micrometres for red light. He predicted that invisible electromagnetic waves ought to exist beyond the visible spectrum in both directions. Within a few decades after he first put forward his theory, physicists had detected and measured the properties of the entire electromagnetic spectrum from gamma and X-radiation at wavelengths less than the diameter of an atom through to radio waves of wavelengths more than 1000 metres. The fact that Maxwell's theory of electromagnetic waves underpins the entire communications industry is a measure of its importance.

Towards the end of the nineteenth century, many scientists held the view that science had little left to reveal in terms of fundamental principles. However, they were unable to explain the continuous nature of the spectrum of thermal radiation emitted by a hot object. The intensity of the radiation peaked at a certain wavelength that depended on the surface temperature of the hot object. The shape of the intensity v wavelength curve could not be explained using Maxwell's theory of electromagnetic waves. The theory could be used to explain why the intensity decreases with increasing wavelength and why it increases with decreasing wavelength. However, it predicted that the intensity becomes infinite, instead of reaching a peak at a certain wavelength which is what was actually measured. The problem became known as the **ultraviolet catastrophe** because it predicted infinite intensity in the ultraviolet part of the electromagnetic spectrum.

The shape of the thermal radiation curve from a hot object preoccupied theoretical physicists for more than a decade at the end of the nineteenth century, until it was satisfactorily explained by Max Planck in Germany in 1900. However, as the basis of his explanation, Planck had to introduce the revolutionary new idea that energy is 'quantized' and is not continuous. He said that the energy of an atom of a hot object could change only by a definite amount or 'quantum' when the atom released or absorbed electromagnetic radiation. Until then, it was generally assumed that energy is a continuous quantity, and that any object can have any amount of energy. Planck explained the shape of the thermal radiation curves by assuming this is not so, and that the atoms of a hot object have well-defined energy levels like the rungs on a ladder. When an atom emits electromagnetic radiation, it moves down a rung. When it absorbs electromagnetic radiation, it moves up a rung.

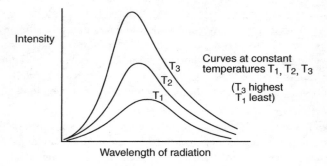

Figure 5.2   Thermal radiation curves

Although Planck showed that the energy of the atoms of a hot object is quantized, he was reluctant to accept that the energy of an electromagnetic wave is also quantized. This was proved by Einstein who used the quantum theory to explain why light would only cause emission of electrons from a cold metal if the light frequency was above a threshold value that depends on the metal. Every atom contains electrons. In a metal, some of the electrons are free to move about inside the metal. However, in a cold metal, the electrons do not have enough energy to leave the metal. They can however obtain sufficient energy from light directed onto the metal – provided the light frequency is above a minimum value.

Photoelectricity was another unexplained discovery made in the closing years of the nineteenth century. It remained unexplained until Einstein put forward the **photon theory of light** in which he said that light consists of packets of electromagnetic waves which he called photons. He said that the energy of a photon is proportional to its frequency. When a metal plate is illuminated with photons of sufficiently high frequency, electrons are emitted from the surface as a result of each electron absorbing a single photon to increase its energy enough to leave the metal. If the light frequency is too low, the energy of each photon is too small to enable an electron to leave the metal. All electromagnetic waves are quantized in the form of photons, which are packets of electromagnetic waves, each carrying well-defined energy in proportion to its frequency. The constant proportionality is the Planck constant, h.

Problems with light

To measure the speed of an object, one method that can be used is to time how long it takes to travel a measured distance. The speed is then calculated by dividing the distance by the time taken. For example, the speed of a cyclist, who travelled a distance of 2000 metres in 100 seconds, would have been 20 m/s. This simple calculation is very deceptive because it assumes distance and time are the same for any observer. Let's look at two situations that reveal some of the problems about speed calculations at speeds approaching the speed of light.

1 Is it possible to catch up with light? The speed of light from a stationary source measured by a stationary observer is 300 000 km/s. One day, travel at speeds approaching the speed of light may well be possible. At the present time, rocket speeds are no more than about 10 km/s. Imagine travelling through space in a rocket at a speed of 100 000 km/s. How fast would a light pulse leave the rocket if it were directed in the same direction as the rocket is moving? Seen from the rocket, it would seem reasonable to predict that the light pulse would move away from the rocket at 200 000 km/s because the light pulse would move 300 000 km in one second and the rocket would move 100 000 km each second in the same direction. The speed of light moving in the same direction as the rocket would be 200 000 km/s. The only problem with this conclusion is that it doesn't agree with the experimental evidence from Michelson and Morley's experiment. The Earth is moving through space at 30 km/s and there is no evidence that this motion affects the speed of light. This conclusion must also apply to the rocket. The speed of light measured on the rocket would be 300 000 km/s, regardless of whether or not it is moving. The distance between the rocket and the light pulse would increase by 300 000 km every second, just the same as if the rocket was stationary. A further question is to consider how the rocket crew could measure their own speed.

2 Is it possible to travel faster than light? In an air display, aircraft sometimes approach each other at high speed, veering away from each other just seconds before a collision. Imagine you are in such an aircraft, flying at 300 metres per second, just less than the speed of sound. The other aircraft is approaching you at the same speed. The two aircraft are closing in on each other at a rate of 600 metres every second. The closing speed is the sum of the individual speeds. This rule was worked out by Galileo in his studies of motion. Apply the same rule to two imaginary spacecraft travelling at 0.6 c, where c is the speed of light in a vacuum. Galileo's rule gives a closing speed of 1.2 c which is more than the speed of light. The rule for adding speeds seems self-evident. Yet is it true at speeds approaching the speed of light? If not, what is the rule for adding such speeds? The answer was provided by Albert Einstein. Now let's look at the ideas that Einstein developed about space, time and speed.

The cosmic speed limit

How fast can you go?

We have seen in the last section that no moving object can catch up with light. The experimental evidence from Michelson and Morley is well established. If nothing can catch up with light, then no two objects can approach each other at speeds in excess of the speed of light. If they could, then a light pulse emitted by one object towards the other would be left behind by the object that emitted it. It seems therefore that no object can be made to travel faster than 300 000 km/s, the speed of light in space.

Imagine a rocket that accelerates away from the Earth at 10 metres per second per second. This acceleration is the same as the acceleration of a freely falling object on the Earth (usually referred to as g). The astronauts on this rocket would feel quite at home as each would experience an inertial force equal to his or her weight on Earth. In practice, rocket accelerations tend to be greater than g. Suppose this acceleration is maintained for a whole year as the rocket travels away from the Earth. Every second, its speed would increase by 10 m/s. After one day, its speed would be 864 km/s, after 30 days more than 25 000 km/s, after a year 315 000 km/s – faster than light! What would happen as the rocket's speed approached the speed of light? It can't exceed the speed of light and it can't catch up with light. Einstein worked out what would happen in his theory of special relativity which he published in 1905. We will look at this in the next few sections.

Einstein, the most remarkable scientist of our age

Do not feel daunted by the scientific facts and theories thus far. Equally, don't be insulted by any simplifications you consider unnecessary. It is well known that Einstein was not a remarkable student. He was born in Germany in 1879 and he and his family moved to Munich in 1885. His work at the Swiss Patent Office in Berne was not too demanding and he would meet his friends in the evenings at the local café. Perhaps because he was free from the direct influence of eminent professors, he started to think about very unusual questions like 'Is it possible to catch up with a light beam?' He worked on his ideas in his free time and his first great paper, *The Electrodynamics of Moving Bodies*, in which he set out his special theory of relativity, was published in 1905. He published two other papers in that year, one of which proved that light consists of wavepackets of energy which he called photons. The other paper explained the haphazard motion of tiny particles observed under a microscope, known as Brownian motion after its eighteenth-century discoverer, Robert Brown. Einstein worked out this motion was caused by repeated and random impacts of gas molecules on each particle. Any of his three papers would have led to the award of the Nobel Prize. He was awarded his PhD in 1905 by the University of Zurich and was appointed to a professorship there in 1909. His academic genius was recognized by fellow scientists and in 1913, he became director of the Kaiser Wilhelm Physical Institute in Berlin. In 1916, he published the *General Theory of Relativity*, in effect his theory of gravity in which he showed that Newton's law of gravitation is a special case of his general theory of relativity. We will look at his ideas on gravity in Chapter 9. Einstein became a celebrity when his ideas on gravity were confirmed by measurements made in the total solar eclipse of 1918. He lectured in many countries, leaving Germany for good in 1933 when the Nazis took control of the country. He emigrated to the United States where he continued his work, attempting to bring quantum theory and his theory of gravity together. His famous remark that 'God does not play dice' shows that he was not well disposed towards the probabilistic nature of quantum theory. He died in 1955, leaving his attempts to unify quantum theory and gravity uncompleted to this day, despite the best efforts of subsequent generations of scientists.

Einstein's theory of special relativity: a layperson's guide

The laws of science are based on objective evidence which produces the same conclusions, regardless of who makes the observations. Einstein realized that if we can't tell if we are moving, the laws of science expressed in the form of equations should be the same for any moving observer. He was not able to do this for every equation using Galileo's rule for adding speeds. One of the main problems was with Maxwell's equations that describe the propagation of electromagnetic waves through space. Einstein used Galileo's rule to change the equations to the coordinate system of a moving observer. The transformed equations should then have been in the same form as the original equations – but they were not and Einstein realized Galileo's transformation rule for adding speeds was the problem. He went back to first principles and set out two postulates or 'starting points':

1 **The speed of light through space is the same, regardless of the motion of the observer or the source**. In other words, no matter how fast you chase after a light beam, it always moves away from you at the same speed.

2 **The laws of physics, expressed as equations, should be in the same form for an observer in any inertial frame of reference**. An inertial frame of reference is defined as a coordinate system in which Newton's laws hold. For example, a train moving at constant speed in a straight line is an inertial frame of reference since an object released by a passenger will behave, relative to the passenger, as if the train were not moving. A train in the process of braking is clearly not an inertial frame of reference! From these two postulates, Einstein showed how to transform equations from any inertial frame of reference to any other one. A detailed analysis is provided in Appendix 2.

Let's consider two of the consequences of Einstein's theory at this stage:

1 A moving clock runs slower than a stationary clock

If two identical clocks are synchronized at the same place (e.g. both set to zero and started at the same place and same time) and one clock is taken off on a high-speed journey then returned, it reads less than the stationary clock on return. Einstein showed that the two readings, t_M for the moving clock and t_s for the stationary clock, are related by the equation

$t_s = \gamma \ t_M$ where γ (pronounced 'gamma') is the **Lorentz factor** $= 1/\sqrt{(1 - V^2/c^2)}$. See page 241 for more details. The Lorentz factor predated Einstein and was the work of Hendrik Lorentz, a Dutch physicist who devised an explanation of the Michelson–Morley experiment in terms of a physical contraction of the apparatus as a result of its motion. Einstein's theory avoids this idea.

For example, for $V = 0.6$ c, then $\gamma = 1.25$ and so $t_s = 1.25 \ t_M$; if the moving clock showed 4.00 on return, the stationary clock would show 5.00.

A tale of two twins

According to the above ideas, if a twin aged 21 took off on a high-speed round trip at Lorentz factor 1.25 and returned 4 years later aged 25, the other twin who stayed on Earth would be 26 when the space traveller twin returned. This might seem very odd, but there is now plenty of scientific evidence to support the theory. The most direct evidence has come from atomic clocks which are remarkably accurate and reliable. An atomic clock flown round the world in a jet plane has a Lorentz factor almost but not equal to 1, just great enough to make a significant difference between it and a stationary clock. Precise measurements on atomic clocks support the theory that a moving clock falls behind a stationary one. A travelling twin would return younger than a stay-at-home twin!

---

2 Moving objects appear to contract

A rod moving at constant speed in a fixed direction appears to be shorter than if it was stationary. Einstein showed that the observed length $L = L_s/\gamma$, where L_s is the length of the rod when stationary (i.e. its proper length). For example, a 1000 metre rocket ship moving at a speed of 0.6 c would appear to be only 800 metres in length because its Lorentz factor of 1.25 would make its observed length equal to (1000/1.25) metres.

Why should this be so?

The time taken for the rocket ship to pass an observer is equal to L/V, its observed length L divided by its speed V. The rocket crew would measure a time for this interval equal to (L_s/V).

Since the crew's clock is stationary relative to the rocket whereas the observer's clock is moving relative to the rocket, then applying $t_s = \gamma \, t_M$ gives $L_s/V = \gamma \, L/V$.

Multiplying both sides of this equation by V gives $L_s = \gamma L$

Hence the observed length of the rocket $L = L_s/\gamma = L_s \sqrt{1 - V^2/c^2}$

The high-speed tube train

Could a high-speed train be trapped in a shorter tunnel if its length is contracted due to its speed? Suppose a 110 m train moves so fast that its length appears to be just 100 m. A Lorentz factor of 1.1 would be needed for this, corresponding to a speed of 0.42 c. Could the train be trapped in a 100 m tunnel? A door operator at the tunnel entrance could be ready to close the tunnel the instant the back end of the train enters the tunnel. The front end would appear to be at the tunnel exit at this instant. However, a light signal from this door operator would take time to reach the door operator at the exit end. The front end of the train would be out by then.

Mass and energy

Faster and heavier – an even stranger result

Einstein's time dilation and length contraction formulas are just two consequences of the theory of special relativity. Yet another mind-bending outcome found by Einstein from his analysis is the discovery that the mass of an object increases the faster it moves. Einstein showed that the mass of a moving object, $m = \gamma m_0$, where m_0 is the rest mass of the object (i.e. its mass measured when it is stationary). According to this formula, if the speed of an object is increased, its mass becomes infinitely large as its speed approaches the speed of light. Einstein went on to show that the kinetic energy of a fast moving object is equal to $(m - m_0) \, c^2$ which led him to his famous formula $E = mc^2$. This formula tells us that if the energy of an object is changed by E, its mass changes by m where $E = mc^2$.

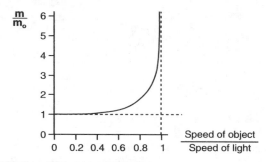

Figure 5.3   Mass v speed

These results astounded Einstein's contemporaries. What evidence was there to support these ideas? Doubt abated when reliable evidence in support of Einstein's formula was produced in experiments on the deflection of high-speed electrons by electric and magnetic fields. More support came from experiments on nuclear physics, where energy unlocked from unstable nuclei could be measured and checked against the mass change of such nuclei. Direct evidence came from high-energy accelerator experiments, all confirming Einstein's mass formulas. At the time of writing, physicists still do not know the mechanism behind $E = mc^2$. What actually causes the mass of an object to change when it is made to go faster? The nature of mass is under intense scrutiny in the latest accelerator experiments, and scientists hope to find out more when the latest high-energy accelerator, the Large Hadron Collider at CERN in Geneva, is completed and put into operation in about 2005.

Matter and antimatter

The existence of antimatter was predicted in 1928 by the British physicist Paul Dirac several years before the first antimatter particle, the positron, was detected. Dirac knew from the work of Lord Rutherford that every atom has a positively charged nucleus surrounded by negatively charged particles called electrons. The nucleus itself is composed of positively charged particles called protons and uncharged particles of about the same mass called neutrons. The rest mass of an electron is much less than the

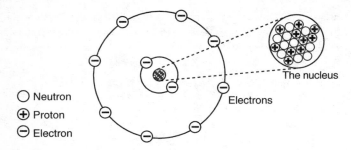

Figure 5.4   The structure of an atom

rest mass of a proton or a neutron. An atom is mostly an empty space, since the diameter of the nucleus is of the order of ten-millionths of the diameter of the atom. However, almost all the mass of an atom is due to the nucleus, so the nucleus is incredibly dense compared with ordinary matter.

■ The lightest atom is the hydrogen atom, which has one electron and a nucleus consisting of one proton.

■ The heaviest naturally occurring atom is the uranium-238 atom which has 92 electrons surrounding a nucleus consisting of 92 protons and 146 neutrons.

Every atom consists of electrons, protons and neutrons. This neat model of atomic structure is sufficient to explain how atoms interact, and how energy is released in chemical and nuclear reactions.

■ A radioactive substance contains atoms with unstable nuclei. An unstable nucleus can emit three types of radiation:

   i)   an alpha particle which consists of two protons and two neutrons. This occurs if the nucleus has too many protons and neutrons

  ii) a beta particle, which is an electron created and emitted from a nucleus when one of its neutrons changes to a proton. This occurs if the nucleus has too many neutrons

 iii) gamma radiation, which is high-energy electromagnetic radiation. This is emitted if an unstable nucleus still has too much energy after it has emitted an alpha or a beta particle.

Where do antimatter particles fit in? The answer is they don't fit into naturally occurring matter. Rutherford and his colleagues carried out experiments over several decades into the properties and nature of naturally occurring radioactive substances. Rutherford's nuclear model of the atom led to the discovery that energy is released when a uranium-235 nucleus (which consists of 92 protons and 143 neutrons) is split as a result of being struck by a slow-moving neutron. This is the principle behind nuclear reactors now in operation on every continent. It is interesting to note that the development of nuclear power stations did not depend on the discovery of antimatter, although our present understanding of **why** energy is released in a nuclear reactor is unlikely to be complete until the nature of matter and antimatter is fully understood. We know how to use Einstein's famous formula $E = mc^2$ to calculate nuclear energy changes, but at the present time, scientists still do not know the fundamental mechanism behind the equation.

Dirac's theory of antimatter

Dirac developed Einstein's energy formula to show that electrons with **negative mass** and positive charge, which he called positrons, ought to exist. He put forward the idea that electrons and positrons are created in pairs from high-energy radiation. The existence of the positron was confirmed several years after Dirac's prediction of its existence. Since then, other types of antimatter particles have been created and detected.

Scientists now know that:

1 there is an antiparticle for every known type of particle, with identical mass and an exact opposite charge
2 a particle and its antiparticle will **annihilate** each other if they meet in a collision, to produce high-energy electromagnetic radiation
3 high-energy electromagnetic radiation is capable of **producing** particle-antiparticle pairs.

A momentous discovery

The existence of the positron was confirmed in 1932 by Carl Anderson, in the United States, who used a device called a cloud chamber in which short-lived visible tracks, like jet trails created by fast-moving charged particles, could be observed. Anderson developed photographic methods to record these tracks. He knew that cosmic radiation crashing into the Earth's upper atmosphere was capable of producing charged particles which created tracks in the cloud chamber. By placing the cloud chamber in a very strong magnetic field, the charged particles could be deflected from their otherwise-straight paths, to produce curved tracks. However, a curved track could be due to a positively charged particle travelling in one direction or a negatively charged particle going in the opposite direction. To find out in which direction a particle was moving, Anderson placed a metal plate vertically in the chamber, reasoning that a particle passing through the plate would lose some energy in the plate and slow down, causing it to be deflected more after passing through the plate. So he could tell from which direction a particle had entered the chamber and could work out from its direction of curvature if it was a positively charged particle, or a negatively charged particle. Most of the tracks he observed were due to electrons, but he found that some were curved in the opposite direction and could only be due to the positively charged electrons predicted by Dirac.

The discovery of the first antimatter particle was described by the Royal Society as the most momentous of the century. Now scientists know that there is an antimatter counterpart or 'antiparticle' for every known particle. However, don't think antimatter is just for scientists, and unlikely to affect you. Positron scanners are now used in hospitals for brain scans, and by aircraft engineers to detect cracks. Sometime in the future, the invention of an antimatter engine could revolutionize space travel, as antimatter fuel could release far more energy than ordinary fuel.

The PET scanner

Positron-emission tomography (PET) scanners are used in hospitals to form brain images. A small quantity of a positron-emitting substance is administered to the patient, whose head is located at the centre of a ring of radiation detectors. Positron-emitting nuclei are created artificially by placing ordinary matter in a beam of high-energy protons. The target nuclei absorb protons and become proton-rich. A nucleus that has too many protons and not enough neutrons is unstable. It becomes stable when one of its protons changes into a neutron. When this occurs, a positron is created and emitted from the nucleus.

The substance reaches the brain *via* the blood system, where it emits positrons. Each positron travels no more than a fraction of a millimetre before it is annihilated by an electron to create two bursts of gamma radiation in opposite directions. This radiation can penetrate ordinary matter easily and triggers two detectors opposite each other. The position of the substance in the brain can be worked out from the detector signals, enabling a computer to build up an image of where the substance is located.

A cosmic problem

Perhaps our part of the Universe is dominated by particles rather than antiparticles. Other parts of the Universe may be dominated by antiparticles rather than particles. Alternatively, perhaps, the Universe overall contains more matter than antimatter. The actual number of 'excess' particles can be estimated, although such an estimate should not be taken as anything other than rough-and-ready figures based on our present incomplete state of knowledge. For example, the Sun is known to contain about 10^{56} particles, mostly protons, since its mass is about 10^{30} kg and there are about 10^{26} protons in each kilogram of matter. The Milky Way is thought to contain about 1 million million stars, giving its total number of particles as 10^{68} (= 1 million million $\times$ 10^{56}). Estimating the number of galaxies in the Universe at about a million million therefore gives about 10^{80} (= 1 million million $\times$ 10^{68}) for the total number of particles in the Universe.

Cosmologists have also estimated that the Universe contains about a billion times as many photons as there are particles. If all these photons could be converted into particle-antiparticle pairs, the Universe would contain one excess particle for every billion particle-antiparticle pairs. However, this estimate assumes all antiparticles have been annihilated, and the Universe consists only of matter particles and radiation. No one knows yet if entire galaxies made of antimatter exist as well as galaxies made only of matter. Perhaps antimatter solar systems exist where there are antimatter humans on antimatter planets. We will return to the story of antimatter in Chapters 12 and 13, when we look at how scientists are attempting to re-create the conditions in the early Universe by means of high-energy accelerator experiments.

Summary

■ **The main parts of the spectrum of electromagnetic waves** (in order of increasing frequency): radio waves; microwaves; infra-red radiation; light; ultraviolet radiation; X and gamma radiation.
■ **A photon** is a quantum of electromagnetic energy in the form of a packet of electromagnetic waves. The energy of a photon is in proportion to the frequency of the waves.
■ **All electromagnetic waves travel at a speed of 300 000 km/s through free space.**

■ **Einstein's Special Theory of Relativity**
  1 The speed of light through space is always the same, regardless of the motion of the light source or the observer.
  2 The laws of physics in the form of equations should be the same in any inertial frame of reference.

Special relativity effects
  1 Time dilation: a moving clock runs slower than a stationary one.
  2 Length contraction: a moving rod appears shorter than a stationary one.
  3 Relativistic mass: the mass of an object becomes ever larger as its speed approaches the speed of light.

■ **Einstein's mass energy equation: $E = mc^2$** where m is the mass of an object and E is the energy equivalent to that mass; c is the speed of light.

■ **Atomic structure:** every atom contains a nucleus which consists of protons and neutrons. The space surrounding the nucleus is occupied by electrons. The hydrogen atom consisting of 1 proton and 1 electron is the simplest atom.

■ **Radioactivity:** a radioactive substance consists of atoms with unstable nuclei. An unstable nucleus becomes stable by emitting either an alpha particle (which is composed of 2 protons and 2 neutrons) or a beta particle (which is an electron created and emitted in the nucleus when a neutron changes to a proton) or a gamma photon.

■ **Antimatter:** for every type of particle, there is an antiparticle which has opposite charge, the same mass and will annihilate its particle counterpart and itself to form high-energy radiation. High-energy radiation is capable of producing particle-antiparticle pairs. The positron is the antimatter counterpart of the electron.

6 | GIANTS AND DWARFS

On a clear dark night, the stars appear as pinpoints of light against a dark background. Before Galileo, the stars were imagined to be fixed to the Celestial Sphere. Now we know that the stars are at different distances from us, ranging from a few light years to hundreds of thousands of light years for the stars in the Milky Way. Galaxies are at different distances too, ranging from a few million light years to thousands of millions of light years. In this chapter, we shall look at how these distances have been worked out. We shall also look at how star masses and diameters have been worked out, how much energy stars emit each second, and how stars are classified and how astronomers can tell if a star is moving towards or away from us. All this information about stars has been deduced by studying the light from stars and using the laws of physics. When you look at a star in the night sky, you can't tell if it is a giant star much bigger than the Sun or if it is a dwarf star, much smaller than the Sun. Accurate measurements need to be made on a star to deduce its exact position, its brightness and the colour of light it emits.

Making measurements

If you were at the Equator, the Pole Star would be exactly on the horizon. The further North an observer is from the Equator, the higher the Pole Star would be in the sky. An observer anywhere North of the Equator can determine his or her latitude by measuring the altitude of the Pole Star above the horizon. For example, an observer who measures 60° for the altitude of the Pole Star would be 60° North of the Equator. It is also possible to work out longitude from astronomical measurements although this method was superseded with the invention of reliable ship-borne clocks or 'chronometers' which enabled local solar time to be compared with the time at a known location. Clearly, the precise measurement of star positions was of great practical importance for navigation purposes before the introduction of modern radar-based and satellite navigation systems.

The success of Newton's theories led astronomers to attempt to measure the distances to stars. It was realized that a star close enough to the Earth should appear to shift its position against the more distant stars over a six-month period as the Earth moves to the opposite side of its orbit round the Sun. This effect is known as **parallax** and may be demonstrated by observing a nearby post against a distant background (e.g. a line of trees or houses): the position of the post against the distant background changes if you move your head to one side. The same effect on a much smaller scale occurs with stars sufficiently close as the Earth moves round its orbit by 180°. However, the shift of position even for the nearest star, Proxima Centauri, is so small that not even the famous eighteenth-century astronomer Sir William Herschel could detect it. To appreciate how small this shift is, a small coin at arm's length makes an angle of less than 1 degree to your eye. The coin would need to be moved to three kilometres away to make an angle to your eye comparable to the shift of position Herschel needed to measure. However, such very small shifts were eventually measured by astronomers in the early nineteenth century. From each such measurement, the distance to the star could be calculated using triangulation, as shown in Figure 6.1.

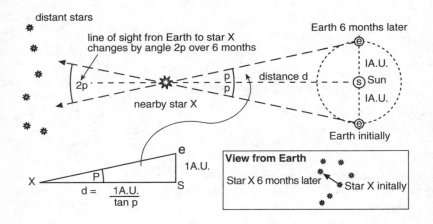

Figure 6.1   The parallax method for measuring the distance to a star

The skinny triangle rule

The triangle formed by any star, the Sun and the Earth is a skinny triangle as it has a very small angle at the star. Very small angles can be expressed in 'minutes of arc' or 'seconds of arc' where 1 degree equals 60 minutes of arc and 1 minute of arc equals 60 seconds of arc. Thus 1 degree = 3600 seconds of arc. Even the nearest star makes an angle of no more than about a second of arc with the Sun and the Earth.

The angle a star makes to the Sun-Earth line is referred to as **the parallax angle** of the star. A star shifts its position by twice this angle when the Earth moves from one side of its orbit to the opposite side.

The mean distance from the Sun to the Earth is defined as 1 astronomical unit (AU). To make an angle of 1 second of arc with the Sun-Earth line, a star would need to be over 206 264 AU from the Sun. The distance from the Sun to a star that makes an exact angle of 1 second of arc with the Sun-Earth line is defined as **1 parsec** (abbreviated as pc).

1 parsec is defined as the distance to a star that makes an angle of exactly 1 second of arc to the Sun-Earth line.

■ The further a star is from the Sun, the smaller the angle it makes with the Sun-Earth line.
At 2 parsecs, it would make an angle of ½ a second of arc.
At 3 parsecs, it would make an angle of ⅓ of a second of arc.
At 4 parsecs, it would make an angle of ¼ of a second of arc.

■ In general terms, the rules of trigonometry can be used to prove the distance formula below for skinny triangles:

$$\text{The distance in parsecs to a star} = \frac{1}{\text{parallax angle in seconds of arc}}$$

■ The parallax method cannot be used beyond a distance of about 50 pc. This is because the parallax angle becomes too small to measure below about 0.020 seconds of arc. The effect of the Earth's atmosphere on light from a star smudges the star's image by about 0.02 seconds of arc. This makes the location of a star to less than 0.02 seconds of arc impossible from the ground. Nevertheless, there are a large number of stars within 50 pc of the Sun, and the parallax method has been used to measure the distance accurately to each such star. These measurements, the first distance measurements beyond the Solar System, have provided the foundation for all other methods of measuring the distances to more distant stars and galaxies.

■ Note that 1 parsec = 206 264.8 AU = 31 million million kilometres = 3.26 light years since 1 AU = 150 million kilometres and 1 light year = 9.5 million million kilometres.

Trigonometry for the terrified

If you want to check that 1 parsec is equal to 206 264.8 AU, don't try a scale diagram as a scale as little as 1 millimetre to represent 1 AU would require a strip of paper over 206 metres long. The trigonometry formula for a right-angle triangle can be used instead. In Figure 6.1, angle X is called the parallax angle.

Using the trigonometry rules for a right-angled triangle, the tangent of angle A, tan A = SE/SX

Rearranging this equation gives SX = SE/tan A.

For angle A = 1 second of arc = 1/3600 of a degree, using a calculator gives tan A = 0.0000048481

Therefore SX = SE/0.0000048481 = 206265 SE

Since the distance SE = 1 AU, then the distance SX = 206 264.8 AU. This is exactly one parsec because angle A was exactly 1 second of arc in this example.

Proper motion

Stars do move through space at very high speeds. Because even the nearest star is so far away, the change of its position relative to neighbouring stars due to its own motion is very small. This is referred to as the **proper motion** of a star. Barnard's star has the largest proper motion of any star at 10 seconds of arc per year. Over the last century, this star has changed its position by about 1000 seconds of arc or one-third of a degree, almost as much as one 'moon width'. To find out if the change of position of a star is due to its own motion (i.e. its proper motion) or the motion of the Earth (i.e. due to parallax), the star's position must be measured three times at six-month intervals apart. The first and third measurements will therefore be made one year apart when the Earth is at the same position. If the star's proper motion is significant, the third measurement will be different from the first measurement. If so, the parallax angle is determined by comparing the midpoint between the first and third positions with the second position.

Magnitude measurements

The brightness of a star depends on how much light the star emits each second and on how far away the star is from us. A very bright star may emit much less light each second than a much fainter star in the same constellation but appears much brighter because it is much closer to us.

Our present system of classifying stars according to brightness is rooted in the ancient world and is thought to have originated with Hipparchus in the third century BC. He divided stars into six categories according to brightness. The brightest stars were referred to as stars of the 'first magnitude' and the faintest, those just visible to the unaided eye, were classified as 'sixth magnitude'. Four further 'magnitudes' were used to classify stars between these limits. This magnitude scale survived through the centuries, even though star brightnesses do not change suddenly from one magnitude to another. A brighter third magnitude star is quite difficult to distinguish by eye from a fainter second magnitude star. Nevertheless, the magnitude scale proves to be a reasonably easy scale to use. For those versed in football, the magnitude scale is not unlike a football league with six divisions, where division one is better than division two which is better than division three and so on. Some stars such as novas even change divisions when they brighten up. A supernova is a star that changes from a low division to a high division suddenly and unexpectedly – this is where the football analogy ends! Star maps often represent star magnitudes by spots of different diameters, with a scale usually given to decode the spot sizes. Don't however think a big dot on a star map represents a giant star. It represents no more than a star that is very bright. The star could be much smaller than the Sun.

Astronomers in the nineteenth century measured the light intensity for stars of differing magnitudes and discovered that a brightness increase by one magnitude (e.g. from 4 to 3) corresponds to about 2½ times as much light, regardless of whether the change is the brighter end or the dimmer end of the magnitude scale. Thus a brightness increase of:

■ two magnitudes corresponds to about 6 (= 2½ × 2½) times as much light,

■ three magnitudes corresponds to about 16 (= 2½ × 2½ × 2½) times as much light,

■ four magnitudes corresponds to about 39 (= 2½ × 2½ × 2½ × 2½) times as much light,

■ five magnitudes corresponds to about 98 (= 2½ × 2½ × 2½× 2½× 2½) times as much light.

This last finding was used to put the magnitude scale on a scientific footing by defining a brightness increase of 5 magnitudes as exactly 100 times as much light. Each brightness increase of 1 magnitude therefore corresponds to 2.512 times as much light since 2.512 × 2.512 × 2.512 × 2.512 × 2.512 is equal to 100.

The classification from first to sixth magnitude is not wide enough to cover all observed star brightnesses at the upper end, and it was extended at the lower end to cover stars that can only be seen through a telescope. By setting m = 6 for the faintest stars visible to the naked eye, telescope stars can be assigned to the magnitude range beyond 6. Very bright stars can be assigned to the magnitude scale beyond m = 1 using the range from 1 to 0 and below 0. Thus, for example, a star of magnitude −1.5 is brighter than a star of magnitude +1.5. On this basis, the Sun's magnitude is about −26 which means the Sun's light is about 150 billion times more intense than a star of magnitude +2 (since 2.512 multiplied by itself 28 times is approximately 150 billion).

Powers for the petrified

The multiplication 2.512 × 2.512 × 2.512 × 2.512 × 2.512 is usually written 2.512^5, where the raised number 5 is the power which 2.512 is raised to.

For two stars, P and Q, of magnitudes m_P and m_Q, where P is brighter than Q,

$$\frac{\text{the light intensity from P}}{\text{the light intensity from Q}} = 2.512^{\Delta m} \quad \text{where } \Delta m = m_Q - m_P$$

This formula can be used for any magnitude difference, regardless of whether or not it is a whole number. The button marked y^x on a calculator is used to raise a number y to a power x. For example, if the magnitude of P is 2.3 and the magnitude of Q is 5.7, then the light from P is 23 times (= $2.512^{(5.7-2.3)}$) more intense than the light from Q.

■ **Light intensity** is the light energy per second falling at right angles onto a surface of area 1 square metre.

Telescopic stars

Galileo was amazed to discover many more stars than anyone had ever seen before when he first used his newly invented telescope to study the night sky. Stars that are too faint to be seen with the unaided eye can be seen using a telescope, or a pair of binoculars, because such instruments collect more light than an eye. This is because the 'collecting area' of a telescope or binoculars is much greater than the eye's pupil, which is the 'collecting area' of the eye. At its widest diameter, the eye's pupil is about 10 mm. A telescope, of diameter 200 mm, has a collecting area that is 400 times greater than that of a pupil, because its diameter is 20 times greater, and the collecting area is proportional to the square of the diameter. Such a telescope could therefore collect 400 times as much light as the unaided eye, corresponding to an extra 6.5 magnitudes (since $2.512^{6.5} \approx 400$). This would allow an observer using this telescope to see 12th magnitude stars. In fact, the largest ground-based telescopes can detect stars fainter than the 27th magnitude, as faint as a torchbulb beyond Pluto.

ACTIVITY

Test a telescope

You will need a telescope or a pair of binoculars for this activity – also a clear night for the second part!

1 Measure the diameter in millimetres of your telescope or binoculars at the entrance. Assuming your eye pupil can widen to 10 millimetres diameter, work out the ratio of the diameter of the telescope to your eye pupil at its widest. Then square this ratio to give the increase of light that can be collected using the instrument. For example, a telescope of diameter 80 mm would enable you to collect 64 times as much light. Then use the magnitude scale in Figure 6.2 to estimate how many more magnitudes can be seen using this instrument. For example, 64 times as much light corresponds to about an extra 4 magnitudes, enabling 10th magnitude stars to be seen on a really clear night.

2 Use a star map to locate and identify a star that is just visible unaided. Then locate and identify another star that is just visible with your instrument. The difference between the magnitudes of the two stars should equal your estimate of how many extra magnitudes your instrument gives.

Absolute magnitude – Big M!

The magnitude of a star, **m**, is a measure of how much light is received from the star. This depends on how much light the star emits and how far away it is. The light from a star spreads out as it travels away from the star, becoming less and less intense further and further from the star. How do you think the Sun's appearance would change to an observer on a spacecraft leaving the Solar System? The solar disc would become smaller and smaller, and the Sun would appear less and less bright to the observer travelling out of the Solar System. At Pluto, a distance of about 40 AU from the Sun, the Sun would appear as an exceedingly bright star, but its radiation would be far too feeble to keep the spacecraft warm. At Proxima Centauri, the nearest star to the Sun at 4.28 light years (= 270 000 AU), the Sun would appear about the same brightness as the Pole Star seen from Earth.

The power or **luminosity** of a star is the energy it radiates into space each second. This is usually expressed in watts, the same as for a light bulb. The Sun emits light at a colossal rate of 400 million million million million watts every second. The luminosity of the Sun is often used as a convenient unit in which to express the luminosity of any other star or a galaxy. For example, the brightest star in the sky, Sirius, is 25 times as luminous as the Sun. Its apparent brightness is because it is relatively close to the Sun, at just 8.7 light years away. In comparison, the second brightest star in the sky, Canopus, in the southern half of the sky, emits 2.5 million times as much light as the Sun. It appears less bright than Sirius though because it is over 130 times further from the Sun than Sirius.

To make a true comparison of the power of different stars, it is necessary to calculate the magnitude each star would have if they were at the same distance from the Solar System. This standard distance is chosen as 10 parsecs for convenience. The magnitude a star would have at this distance is referred to as the **absolute magnitude M** of the star. This can be calculated from the magnitude m of the star if its distance is known. The calculation is based on the principle that the intensity of light at a certain distance from a point source varies with the inverse of the square of the distance. In other words, if the light intensity at a certain distance d_0 is I_0, then the light intensity at

double this distance (i.e. $2d_0$) is ¼ I_0,
treble this distance (i.e. $3d_0$) is ⅑ I_0,
four times this distance (i.e. $4d_0$) is ¹⁄₁₆ I_0, etc.

The inverse square law for light from a point source follows from the geometrical rule that the area of a sphere is proportional to the square of the sphere's radius. Imagine a point source at the centre of a sphere. Each square centimetre of the sphere's surface would receive a certain amount of light. If the sphere were doubled in diameter, its surface area would be four times greater so each square centimetre of this expanded surface would receive one quarter as much light as it previously received.

To see how the inverse square law works for a star, suppose Sirius (m = −1.4) was moved from its present position which is 2.7 parsecs (= 8.7 light years) from the Sun to a distance of 10 parsecs from the Sun. As the distance to Sirius would therefore increase by a factor of 3.7 (= 10/2.7), the amount of light received from Sirius would decrease by a factor of 13.7 (= 3.7 × 3.7) because of the inverse square law. How many magnitudes does a factor of 13.7 correspond to? You could use Figure 6.2 to work this out. You should obtain an answer of 2.8. The magnitude of Sirius would therefore be +1.4 at 10 parsecs. In other words, the absolute magnitude of Sirius is +1.4.

The absolute magnitude M of a star is the magnitude it would have if it was at a distance of exactly 10 parsecs from Earth.

A ready reckoner The method of calculating the absolute magnitude M of a star from its apparent magnitude m and its distance has been converted into a conversion graph in Figure 6.2. The point on the line for any given distance d in parsecs on the horizontal axis is located. The magnitude difference or **distance modulus** m − M for this point corresponding to d is then read off the vertical axis. The absolute magnitude M can then be calculated by subtracting m − M from the magnitude m.

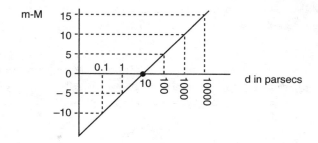

Figure 6.2   A ready reckoner

ACTIVITY

A calculation on Canopus Canopus, the second brightest star in the sky, has a magnitude of –0.7. This star lies at a distance of 360 parsecs (= 1170 light years) from the Sun. Prove for yourself that if Canopus were moved to 10 parsecs from the Sun, we would receive 1300 times more light from it. Hence show that the absolute magnitude of Canopus is –8.5.

Contrasts and comparisons

The colour of a star is far from evident when it is viewed unaided, except for very bright stars like Sirius. Star colours are much more obvious when a telescope is used. This is because the colour-sensitive light cells of the retina of the eye do not work at low light intensities. A different type of retinal cell which is not colour-sensitive functions at low light intensities which is why we cannot tell the colour of a very dim object. When a star is observed using a telescope, much more light is collected by the telescope than by an unaided eye, so the colour-sensitive cells are more likely to be stimulated when a telescope is used.

Stars vary in colour, covering a spectrum that includes red, orange, yellow, white and bluish stars. The present system of classifying stars was originated by astronomers at Harvard in about 1900. The system makes use of the fact that the colour of a star depends on its surface temperature. This link between colour and temperature can be demonstrated by observing how the colour of torchbulb filament changes as the filament current is increased. The filament changes colour from dull red to orange-red to yellow as it becomes hotter. Clearly a red star such as Betelgeuse is not as hot as a bluish-white star such as Rigel. The Harvard astronomers assigned a classification letter to each distinctive star colour and were able to relate each colour to a temperature range. Figure 5.2 shows how the spectrum of thermal radiation from a hot body depends on the surface temperature of the body. By measuring the wavelength of light at which a star emits its peak intensity, the surface temperature of the star can be deduced.

The Harvard system was developed and continues to be used as the standard classification system of stars. The letters were carried forward from a previous classification system which is why they are not in alphabetical order as the temperature changes.

Star type	O	B	A	F	G	K	M
Colour	blue	bluish-white	white	yellow	yellow-orange	orange-red	red
Temperature /kelvins	30 000	20 000	10 000	8,000	6,000	4,000	3,000

Notes

1 The kelvin (K) is the scientific unit of temperature, defined by assigning zero to the lowest possible temperature and 273 K for the temperature at which ice, water and water vapour are in thermal equilibrium. Measurements of the thermal radiation from hot objects in the laboratory proved that the wavelength at peak intensity $\times$ the surface temperature is a constant (= 0.0029 K m). The Sun's radiation peaks at 0.0005 mm, corresponding to a surface temperature of 5800 K.

2 Numerical sub-classes within each main class were established by the Harvard astronomers on the basis of the detailed spectrum of light from each type of star.

3 The order of the letters can be recalled using the mnemonic 'Oh Be A Fine Girl/Guy Kiss Me'

The Hertzsprung–Russell diagram

This diagram was devised independently by the Danish astronomer Enjar Hertzsprung in 1911 and the US astronomer Henry Russell in 1913. Each star is plotted on the diagram according to its absolute magnitude and its temperature. The result is shown in Figure 6.3.

■ Stars range in absolute magnitudes from about +15 which is 10 000 times less powerful than the Sun, to about −10 which is a million times more powerful than the Sun.

■ Most of the stars on the diagram lie on a diagonal belt which runs from the bottom right-hand corner to the top left-hand corner. This is referred to as the **Main Sequence**. The Sun at M = +5 and a temperature of about 6000 K lies near the middle of the Main Sequence.

■ The very powerful M-class stars high above the Main Sequence are referred to as **giants** or **supergiants**. These stars are much larger than the Sun which is why they are collectively so called. This can be deduced from the fact that an M-class star is cooler than the Sun so it emits less light per unit area of its surface. It must therefore be much larger than the Sun if it is higher up the HR diagram than the Sun as it emits more power. Giant stars lie about 5 magnitudes above the Sun on the HR diagram. Supergiants lie about 5 magnitudes higher than giant stars.

■ The group of very hot faint stars below the Main Sequence are referred to as **white dwarfs**. The surface temperatures of these stars are much higher than that of the Sun so a white dwarf emits much more light per unit area than the Sun does. However, compared with the Sun, a white dwarf is a feeble emitter of light so it must be much smaller in diameter than the Sun.

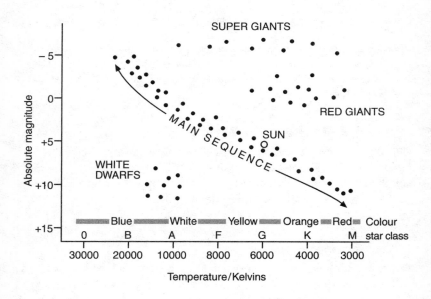

Figure 6.3   The Hertzsprung–Russell diagram

We will meet the HR diagram several times again in the later parts of this book. In Chapter 7, we will look at how stars evolve from birth to death, and how giants and dwarfs form part of this lifecycle. In Chapter 10, we will look at how the HR diagram has been used to deduce the age of the oldest stars in the Milky Way. Problems arose for cosmologists as a result of estimates of the age of the Universe which turned out to be less than the age of these very old stars – older than the Universe itself!

ACTIVITY

Hunting for giants and dwarfs

1 Arcturus is an orange-red star of magnitude –0.1 in the constellation of Bootes the Herdsman at a distance of 37 light years (= 11.3 pc) from the Sun. It can be seen by observers in the Northern hemisphere due South before midnight in May. Its absolute magnitude is –0.4 and its colour makes it a K-class star, cooler than the Sun. However, since it emits more light than the Sun, it must be larger than the Sun. In fact, it is 40 times wider than the Sun. It is therefore classed as a giant star. Supergiants lie above the giant stars on the Hertzsprung–Russell diagram.

2 Antares is an M-class supergiant in the constellation of the Scorpion. It can be seen due South at about midnight in May and June by an observer in the Northern hemisphere but much nearer the horizon than Arcturus. At magnitude 0.9, Antares lies 520 light years (= 160 parsecs) from the Sun. Its absolute magnitude at –5.1 puts it significantly above the giant stars on the HR diagram. In fact, it is 300 times wider than the Sun, large enough to take in the orbit of Mars.

3 Sirius B is an A-class star of magnitude 8.7 in orbit about Sirius in the constellation Canis Major – the Greater Dog. To an observer in the Northern hemisphere, Sirius lies due South at midnight in late December. A large telescope is needed to see Sirius B, as it never strays more than about 10 seconds of arc from its much brighter parent, Sirius, as they move round each other in a 50-year orbit. At 8.7 light years (= 2.7 pc) from the Sun, Sirius B has an absolute magnitude of about +11 which puts it well below the Main Sequence on the HR diagram. Although its mass is about the same as that of the Sun, its diameter is only about one fiftieth of the Sun's diameter.

The laws of physics at work – again!

The amount of light emitted per unit area from the surface of a hot object varies according to the fourth power of its surface temperature. This law was discovered as a result of laboratory experiments, and it underpins all the estimates of star diameters. For example, the surface temperature of Arcturus deduced from its colour is known to be 3100 K. Since the surface temperature of the Sun is known to be 5800 K, the surface temperature of Arcturus is therefore 0.53 × the surface temperature of the Sun. According to the fourth power rule, Arcturus therefore emits 0.08 times (= 0.53 × 0.53 × 0.53 × 0.53) as much light per unit area as the Sun does. However, the absolute magnitude of Arcturus is –0.6 whereas the Sun's absolute magnitude is +4.8. Therefore, Arcturus emits 145 times as much light as the Sun does (since its absolute magnitude differs by 5.4 and $2.512^{5.4} = 145$; see page 71 if necessary). Arcturus must therefore have a surface area which is over 1800 times greater (= 145/0.08) than that of the Sun. Its diameter must therefore be over 40 times greater than that of the Sun. This example shows how the fourth power law is used to deduce the diameter of a star in comparison with the Sun.

Star spectra

A rainbow is a dramatic sight, caused by raindrops splitting sunlight into the colours of the spectrum. Sir Isaac Newton discovered how to split sunlight into a spectrum, using a glass prism. He placed the prism in the path of a narrow beam of sunlight which entered a darkened room through a small hole in the window blinds. The beam was refracted by the prism and split into colours, which Newton saw on the opposite wall of his chamber. A light spectrum can also be produced by directing a narrow beam of white light from a suitably enclosed lamp bulb through a prism. The spectrum is referred to as a white light spectrum, since it is produced by splitting a beam of white light into a continuous spread of colours.

The spectrum of sunlight and of a light bulb are both examples of **continuous light spectra** where the colours on the screen change continuously with position from red through the colours of the rainbow to blue and violet, as below.

Red   Orange   Yellow   Green   Blue   Indigo   Violet

In fact, the spectrum of sunlight is much more extensive at the blue end than the spectrum of a light bulb. The reason why very hot stars are bluish-white, rather than white, is that very hot stars produce even more blue light than the Sun as well as all the other colours of the spectrum.

A light beam can also be split into a spectrum using a device called a **diffraction grating**. This device consists of a glass plate on which many evenly spaced parallel slits are ruled. The effect of this device on a parallel beam of light of a single colour is to split the light in certain directions only. The angle which each beam makes with the original beam is measured and used to calculate the wavelength of the light beam from a standard formula, provided the slit spacing is known. With a white light beam, each colour component can be measured to give its wavelength.

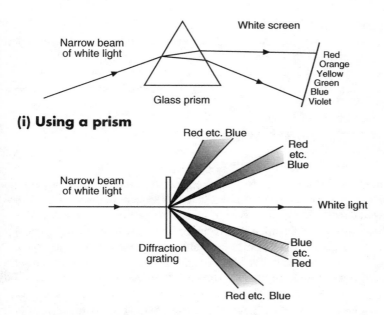

(i) Using a prism

(ii) Using a diffraction grating

Figure 6.4(a)   Producing a continuous spectrum

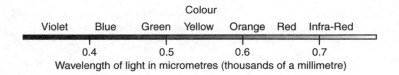

Figure 6.4(b)   Wavelength and colour

Line spectra

The solar spectrum is crossed by thin black vertical lines. These were first observed and investigated in 1817 by the German physicist Joseph Fraunhofer. He labelled these lines alphabetically from the deep red end of the spectrum, and he measured the wavelength of each line using a diffraction grating. The cause of these lines, which became known as Fraunhofer lines, was a puzzle to Fraunhofer. He knew that they were caused within the Sun, and were not due to the Earth's atmosphere. The detailed explanation was not forthcoming for many years, even though Fraunhofer and other astronomers observed similar but not identical patterns of black lines in the light spectra from other stars.

The study of light spectra from ordinary light sources and from stars became a major activity for Fraunhofer and his contemporaries. They discovered that coloured flames are produced by putting certain metal salts in a flame, for example calcium salts make the flame turn red and sodium salts (e.g. common salt) make it turn orange. The spectrum of light from a coloured flame was found to contain very strong coloured lines. These coloured lines are referred to as a **line emission spectrum**. Such spectra can be much more easily observed now using vapour lamps instead of coloured flames, but the early investigations were before such lamps were invented. Using a vapour lamp, a pattern of discrete coloured lines against a dark background is observed. The pattern of the lines is characteristic of the chemical elements that produce the light. This was first established by Gustav Kirchhoff who proved that observations of spectra can be used to identify the elements present in coloured flames. For example, the sodium spectrum includes two closely spaced prominent yellow lines at wavelengths of 580.0 and 580.6 nanometres.

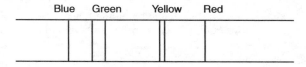

Figure 6.4(c)   A line spectrum

Kirchhoff made an even more important discovery in 1859 when he established that black lines are observed against the white light spectrum when white light from a laboratory light source is passed through flames containing common salt which is a compound of sodium and chlorine. Without the flame in the path of the white light beam, a continuous spectrum is produced without any black lines. When a 'sodium' flame is placed in the path of the light beam, black lines like the Fraunhofer lines of the Solar Spectrum appear. The pattern of black lines is called a **line absorption spectrum**. Kirchhoff discovered that the line absorption spectrum of sodium includes two prominent black lines at 580.0 and 580.6 nanometres – exactly the same position as the two yellow lines in the line emission spectrum of sodium. He knew that white light covers a continuous spread of colour, from blue at about 400 nanometres to red at about 650 nanometres. He deduced that the sodium flame absorbed light of wavelengths 580.0 and 580.6 nanometres when a white light beam was passed through it, causing the light beam to continue without light at these two wavelengths. Kirchhoff recognized that the same process explains the Fraunhofer lines of the Solar Spectrum. The Sun is surrounded by very hot gases which are only visible as the solar 'corona' when a solar eclipse occurs and the solar disc is blocked out by the Moon. Kirchhoff realized that white light from the Sun's photosphere, its light-emitting surface, passes through the surrounding hot gases which absorb certain wavelengths, depending on the elements present in the hot gases. These wavelengths are therefore not present in sunlight and appear as black lines against the continuous spectrum of sunlight. By measuring the wavelengths of these black lines and comparing these measurements with the wavelengths in the line emission spectra of all the known elements, Kirchhoff was able to identify the chemical elements present in the solar corona. Even more importantly, he realized that the chemical elements in any star could be identified by observing the spectrum of light from the star and measuring the wavelengths of the absorption lines in its spectrum.

The chemistry of the stars

Kirchhoff's work was developed by other scientists and astronomers who found an abundance of hydrogen in many stars as well as many metal elements, metal oxides and carbon. In 1868, the English astronomer Norman Lockyer discovered absorption lines in the solar spectrum that did not correspond to any known element. He concluded that they were due to a new element which became known as helium after 'helios' – the Greek word for the Sun. The existence of this new element was not however confirmed until 1894 when it was discovered and identified in gas from underground wells. In fact, helium is the second lightest element and is chemically inert. The helium gas trapped underground has been there since the formation of the Solar System. Lockyer's discovery was followed by confirmation that helium is present in many other stars. We shall see in Chapter 7 that the abundance of helium in the stars is because protons are fused in a nuclear reaction to form helium nuclei inside stars in a process that releases energy. We shall also see in Chapter 10 that the measured ratio of helium to hydrogen is an important consideration in the evidence for the Big Bang.

The most detailed study of the line absorption spectra of stars was carried out by E.C. Pickering at Harvard in about 1900. He and his colleagues devised the Harvard classification system outlined on page 73 on the basis of the elements present as well as the surface temperature. Absorption lines were observed due to molecules and to ions which are charged atoms as well as due to neutral atoms. Their observations are summarized in the table opposite.

Star type	O	B	A	F	G	K	M
Temperature K (kelvins)	30 000	20 000	10 000	8000	6000	4000	3000
Most significant elements present	Helium ions	Helium and hydrogen atoms	Hydrogen atoms; calcium, magnesium and silicon ions	Calcium ions; iron, sodium atoms and ions	Calcium ions and atoms; other metal atoms	Metal atoms	Molecules (e.g. titanium oxide); metal atoms

Notice that hydrogen and helium are not significant at temperatures below about 10 000 K. At these temperatures, metal atoms and ions are dominant. Molecules are present in the coolest stars. Clearly, the discovery of different chemical elements at different temperatures gave astronomers the enormous task of explaining why these elements are present and why some are only present in very hot stars whereas others are present only in much cooler stars.

The Doppler effect

So far, we have seen that the light from a star can be used to deduce its temperature, its light output, its diameter and its chemical composition. In addition, the speed at which the star is moving away from or towards the Earth can be deduced. This is because the wavelength of light emitted by a star moving away from or towards us differs from what it would be if the star were not moving. The change of wavelength due to the motion of the source is called the **Doppler effect**, after Christian J. Doppler who first put forward the idea in a paper on double stars in 1842. At the time, wavelength measurements were not sufficiently accurate to confirm Doppler's ideas. However, a few years later, the effect was demonstrated with sound waves by making a railway truck carrying a musician playing a note of constant pitch move rapidly past a line of people with 'trained' ears. Each listener heard the pitch of the note rise then fall as the truck approached, went past and then receded into the distance. The same effect can be heard by a pedestrian when an emergency vehicle sounding a high-pitched siren rushes past.

The pitch of a sound note is a measure of the frequency of the sound waves, so a rising pitch on approach corresponds to a shortened wavelength. A falling pitch on receding corresponds to an increased wavelength. Figure 6.5 shows why this change of wavelength occurs. The source emits waves of constant frequency that spread out from the source in all directions.

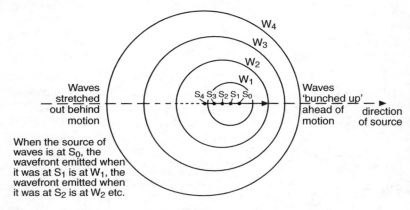

Figure 6.5   The Doppler effect

- The wavefronts in the direction of motion of the source are closer together because the source is moving in the same direction as the waves. The wavelength is shortened by an amount equal to the distance moved by the source each cycle.
- The wavefronts in the opposite direction to that of the source are further apart because the source is moving in the opposite direction to the waves. The wavelength is lengthened by an amount equal to the distance moved by the source each cycle.

With light from an approaching or receding star, the change of wavelength causes all the absorption lines to shift their positions in the same direction. The pattern remains unchanged but the position of every line is shifted in the same direction. The light is said to be:

1 **blue shifted** if the source is approaching so the wavelength is shortened. This is because blue light has a shorter wavelength than red light.
2 **red shifted** if the source is receding so the wavelength is lengthened.

The change of wavelength, usually written $\Delta\lambda$ (pronounced 'delta lambda') increases with the speed at which the star is approaching or receding from us. By measuring the change of wavelength, this speed can be worked out. The following formula explained below is used for this.

Speed of approach or recession of star =

$$\frac{\text{change of wavelength}}{\text{wavelength from a stationary source}} \times \text{speed of light}$$

For example, if the spectrum of light from a star shows the absorption lines shifted to longer wavelengths, the star must be receding from the Earth. If one of these lines is shifted by 5.0 nm from 580.0 nm to 585.0 nm, the star's speed of recession = $\frac{5}{580} \times 300\,000$ km/s ≈ 2600 km/s.

The Doppler effect

The change of wavelength $\Delta\lambda$ increases with the speed υ of approach or of recession of the source in accordance with the equation $\Delta\lambda = \frac{\upsilon}{c}\lambda$, where c is the speed of the waves.

This is because the source moves a distance $\Delta\lambda$ equal to υ t in the time (t) it takes to emit one cycle. Since the wave travels a distance λ in this time, then $\lambda = ct$.

Hence $\Delta\lambda = \upsilon\, t = \upsilon\,\dfrac{\lambda}{c}$

Rearranging this formula gives $\upsilon = \dfrac{\Delta\lambda}{\lambda}\, c$

Summary

Star positions

■ **Parallax** is the change of apparent position of a nearby object relative to distant objects due to the movement of the observer.
■ **The parallax angle** of a star is the angle between the line from the star to the Sun and the line from the star to the Earth.
■ **The parsec** is the distance to a star that makes an angle of exactly one second of arc with the Sun and the Earth.
■ **1 degree** = 60 minutes of arc = 3600 seconds of arc

■ **Distance in parsecs to a star** =

$$\frac{1}{\text{parallax angle in seconds of arc}}$$

■ **Proper motion** is the movement of a star across the line of sight

Stars and brightness

■ **Magnitude change**

Magnitude change	1	2	3	4	5
Change of light received	×2.5	×6	×16	×40	×100

■ **Luminosity or power** of a star is the energy it radiates into space each second
■ **Absolute magnitude M** of a star is its magnitude if it was at a distance of 10 parsecs from us
■ **Star classification**

Star classification	O	B	A	F	G	K	M
Temperature/kelvin	30 000	20 000	10 000	8000	6000	4000	3000

■ **A red giant** is a K or M class star that is much larger than the Sun.
■ **A white dwarf** is an O or B class star that is much smaller than the Sun.

Spectra

■ **A continuous spectrum** includes all the colours of the spectrum
■ **A line emission spectrum** consists of coloured lines at definite wavelengths
■ **A line absorption spectrum** consists of black lines at definite wavelengths on a continuous spectrum

The Doppler effect

■ **A blue shift** is a decrease of wavelength due to a star moving towards the observer
■ **A red shift** is an increase of wavelength due to a star moving away from the observer
■ **The speed** at which a star is approaching or receding is given by

$$\text{speed of star} = \frac{\text{change of wavelength}}{\text{wavelength from a stationary source}} \times \text{speed of light}$$

7 | THE LIFE OF A STAR

The Sun is thought to be about 5000 million years old, about half-way through its life as a star. Stars vary in age from a few million years or less to over ten thousand million years. So far, we have seen how the luminosity of a star is determined and how star diameters can be worked out. In this chapter, we will look at how the mass of a star is determined, and why this enables astronomers to work out how old a star is. We shall see what changes take place after a star forms, how energy is produced and released in a star and what happens to a star when its source of energy runs out. Massive stars end their lives in a huge 'supernova' explosion that can outshine an entire galaxy for weeks. We shall meet supernovae in the next chapter where we will see how they are used to measure distances to galaxies. Distant supernovae are important events that tell astronomers about the conditions in the early Universe. In this chapter, you will also find out what astronomers expect to happen to the Sun when it runs out of energy. There is good evidence that it has happened before but next time could be different.

Binary stars

Imagine living in a solar system with two suns, in orbit about each other. Perhaps one of the two suns would always be in the sky of a planet in this solar system. Without night time, anyone on such a planet might never see the night sky and never know about other stars. Systems consisting of two or more stars in orbit about each other are more common than you might think. For example, the next time you observe the Plough in Ursa Major with a telescope or with binoculars, look closely at Mizar and its fainter companion Alcor which are located adjacent to Benetash which is at the end of the 'handle'. Mizar and Alcor are about 27 parsecs from us, separated by a distance of less than 0.1 parsecs. Mizar and Alcor are an example of binary stars, in orbit about each other. They can just be seen separately without the aid of a telescope or binoculars if you have very sharp eyes. In fact, observed through a telescope, Mizar itself can be seen as a binary or double star, its two components being about 14 seconds of arc apart. This discovery was made in 1650. Astronomers now know that each of the two components of Mizar are also double stars and that Alcor is also a double star – six stars in a complicated binary system. The pattern of day and night would be very complicated on a planet in this system.

A catalogue of double stars was published by Sir William Herschel in 1782. He thought that double stars were chance alignments of pairs of nearby and distant stars. He thought that all stars have the same luminosity, so where one star is brighter than the other in a double star, the brighter star must be nearer to us. However, his attempts to measure parallax between the two stars of a double star were not successful. As a result of making measurements on a number of double stars over a period of years, he concluded that these double stars were orbiting round each other and that stars do differ in luminosity. Most double stars are binary stars in orbit round each other. A **visual binary** is a pair of stars that are observed to be in orbit round each other. Many binary stars never appear as any more than a single point. If the two stars eclipse each other as seen from the Earth, the overall brightness dips at each eclipse thereby revealing the 'star' as a binary. A more important indicator of a non-visual binary is a periodic shift of the wavelength of the lines of the light spectrum. Such stars are too far away to resolve visually and they are referred to as **spectroscopic binaries**.

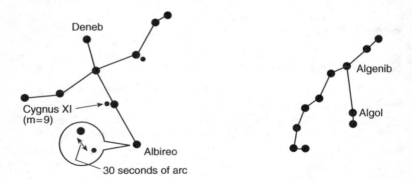

Figure 7.1   Searching for double stars

ACTIVITY

Detecting double stars

Using a telescope or binoculars on a tripod or a similar mount, see if you can make the following observations:

1 **The two components of Mizar**. You won't be able to observe the two components of Alcor or the binaries of the individual components of Mizar because they are too close to resolve. Astronomers know that these stars are doubles from their spectra, as explained later in this chapter.

2 **Albireo**, a double star about 125 parsecs from us, is at the 'tail-end' of the constellation Cygnus. It is prominent in the night sky in summer in the Northern hemisphere. The two components are of different colours and are separated by more than 30 seconds of arc.

3 **Algol**, an eclipsing binary star, lies in the constellation Perseus, visible in the night sky for most of the year in the Northern hemisphere. Its magnitude varies between 2.1 and 3.4 and back once every two and a half days. It was known to ancient astronomers as the Demon star, who imagined it was blinking. In fact, its variable magnitude is because it has two components which regularly eclipse each other. The components are unequal in brightness and about the same diameter, so when the fainter component eclipses its companion, the overall brightness of Algol drops then rises back. At less than 0.1 AU apart, the two components are too close to resolve using even the most powerful telescope.

Spectroscopic binaries

The lines of the absorption spectrum of a star are blue shifted if the star is approaching and red shifted if the star is receding. Two stars in orbit about each other repeatedly move towards and away from us as they move round each other. Only if the plane of the orbit is perpendicular to the line of sight from the Earth is the distance to each star constant. Each absorption line splits open and closes again repeatedly if the light is from a binary star in which the two stars are repeatedly moving towards and away from us. Figure 7.2 shows the idea. The light from star X is blue shifted as it approaches then red shifted as it recedes. Star Y is always moving in the opposite direction to X so when the light from X is blue shifted, the light from Y is red shifted and when the light from X is red shifted, the light from Y is blue shifted.

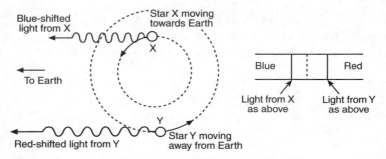

Figure 7.2   A spectroscopic binary system

■ If the lines split symmetrically, the two stars must be moving round their exact mid-point at equal speeds in opposite directions. The two stars must therefore have equal mass otherwise they would orbit about a point nearer the more massive star.

■ If the lines do not split symmetrically, the two stars must be moving round at different speeds since they produce unequal wavelength shifts. The centre of rotation is not therefore at the mid-point and is nearer the more massive star which has a smaller orbit than the other star.

The time taken for the stars to complete one orbit, the time period, is measured directly from the time taken for each spectral line to open and close again twice. The time period can be measured directly for visual binaries.

The speed of each star can be worked out by measuring the maximum change of wavelength and using the equation for the Doppler effect on page 83.

The radius of orbit of each star can then be calculated from the fact that the circumference of the orbit (= $2\pi \times$ the radius) is equal to the speed of the star multiplied by its time period. Adding the orbital radius of one star to the orbital radius for the other then gives their separation.

The separation of the two stars can be worked out for visual binaries by triangulation if the maximum angle between the two stars is measured and if the distance to the two stars is known.

Star masses

If the time period and separation of the two stars in a binary system are known, the mass of the stars can be worked out using the form of Kepler's 3rd Law stated below. The theory behind this equation is explained on page 239. By expressing the time period in years and the separation in astronomical units, the mass of the two stars in the equation is in solar masses.

$$\text{Total mass} \times (\text{Time period})^2 = (\text{Separation})^3$$
$$\text{(in solar masses)} \qquad \text{(in years)} \qquad \text{(in AU)}$$

The mass of each star can then be worked out if the radius of orbit of each star is known because the mass ratio of the two stars is the inverse of the ratio of their radii of orbit.

Using this method, astronomers worked out that star masses lie roughly in the range from about one tenth of the mass of the Sun to about fifty times the mass of the Sun. We shall see later in this chapter why a minimum amount of mass is necessary for an object to become a star and emit light. If Jupiter had been ten times as massive, although still much less massive than the Sun, it would have become a star. We shall also see why high-mass stars are much shorter-lived than the Sun.

The mass of the Sun can also be worked out using the theory explained on page 239. The mean radius of the Earth's orbit is 1.5×10^{11} m and the Earth takes 32 million seconds (= 365¼ days) to move round its orbit once. Substituting these values into the equation on page 239 together with the value of G measured in the laboratory, the Sun's mass works out at 2.0 million million million million million (= 2.0×10^{30}) kg.

Worked example

Two stars in a binary system orbit each other once every 50 years on orbits of radii 20 AU and 5 AU. Calculate the mass of each star in the system.

Solution

Separation = 20 + 5 = 25 AU

Time period = 50 years

Using the equation **Total mass $\times$ (Time period)^2 = (Separation)^3**

Total mass $\times 50^2 = 25^3$

$\therefore$ total mass $= \dfrac{25 \times 25 \times 25}{50 \times 50} = \dfrac{15625}{2500} = 6.25$ solar masses

Since the ratio of radii = 20/5 = 4, the ratio of star masses = ¼

$\therefore$ The mass of one star = 1.25 solar masses and the mass of the other star = 5.0 solar masses.

Luminosity and mass

The luminosity of a star is a measure of its light output. We saw in Chapter 6 how this can be determined if the apparent magnitude and the distance to the star are known. In the previous section of this chapter, we saw how the mass of a binary star can be worked out. If a binary star is close enough to us, its luminosity and its mass can be determined separately. Both of these quantities are usually expressed relative to the Sun.

By 1920, Sir Arthur Eddington had gathered enough data on binary stars to show that the higher the mass of a star, the greater its light output is. For **main sequence** stars, star masses vary from about 0.1 solar masses at the foot of the main sequence where stars are less than one ten thousandth as luminous as the Sun, to about 30 solar masses at the top where the stars are over a million times as luminous. Eddington showed that for main sequence stars, the luminosity is approximately proportional to the cube of the mass. In other words, a star that has

- a mass twice that of the Sun emits approximately 8 times ($= 2 \times 2 \times 2$) as much light as the Sun,
- a mass three times that of the Sun emits about 27 times ($= 3 \times 3 \times 3$) as much light as the Sun,
- a mass ten times that of the Sun emits about 1000 times ($= 10 \times 10 \times 10$) as much light as the Sun.

Eddington's luminosity v mass relationship can be used to find the luminosity and hence the absolute magnitude of binary stars of known mass at unknown distances. Beyond about 200 parsecs, the parallax method of measuring distance is not possible as the parallax angle is too small. Binary stars beyond this distance can nevertheless be studied spectroscopically to enable their masses to be found. Knowing the mass of a binary star enables its luminosity and hence its absolute magnitude to be estimated if it is a main sequence star. Its distance can be estimated using the relationship between absolute magnitude, distance and observed magnitude, as explained on page 70. In addition to enabling binary distances to be worked out beyond 200 pc, the luminosity v mass relationship provides a target for theoreticians to explain. Any model of the processes and structure of a star must produce this luminosity v mass relationship.

Inside a star

Stars appear as pinpoints of light, steadfastly releasing energy on a colossal scale. The nearest star to us, the Sun, releases energy at a rate of about 400 million million million million watts. A very large power station produces no more than 10 000 million watts, enough to meet the energy needs of everyone in a large city. Large power stations produce electricity as a result of burning fossil fuels such as coal, oil or gas or as a result of nuclear fission. Where does the Sun obtain its energy from?

More about sunlight

A light bulb emits light from its filament. The filament is a metal wire heated by an electric current to such a high temperature that it glows. Street lights are usually vapour lamps that emit light as a result of a metal vapour in the lamp bulb being excited by an electric discharge. Such lamps emit light of a characteristic colour. For example, yellow light is produced by a sodium vapour lamp. The characteristic colour is because the electrons in the vapour atoms are at specific energy levels. The atoms repeatedly collide with each other. In a collision between two atoms, some of the electrons in each atom might hop to higher energy levels away from the nucleus. This 'excited' state of an atom is unstable and temporary, ending when the electrons fall back to a lower level. When an electron makes such a move, it releases energy as a photon of light. The photon energies and wavelengths of light from a vapour lamp depend on the energy levels of the electrons round the nucleus of the atom. This is why the light from a vapour lamp gives a line spectrum. In a filament lamp, the atoms vibrate about fixed positions so the energy levels become energy bands, allowing electrons to release photons with a continuous range of energies and hence a continuous spectrum.

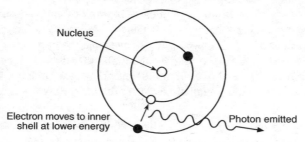

Figure 7.3   A photon being emitted from an atom

The same process of light emission as in a filament lamp is at work in the Sun's **photosphere**, the light-emitting outer layer of the Sun that we see as the solar disc. The Sun is a glowing ball of exceedingly hot gas, mostly consisting of hydrogen and helium atoms without electrons. The temperature at the centre of the Sun is thought to exceed ten million degrees and its internal pressure, due to its own weight, is thought to be over five billion times higher than that of the Earth's atmosphere. The temperature inside the Sun decreases towards its edge. At the photosphere, electrons and nuclei move slowly enough to form ions which are charged atoms. A continuous spectrum of light is emitted by electrons in this layer just as in a filament lamp. The photosphere defines what we see as the solar disc. It is sometimes referred to as the surface of the Sun although it is not like the surface of a liquid or a solid. Above the photosphere, a thin layer of gaseous elements as heavy as iron absorb light photons emitted from the photosphere, causing absorption lines in the solar spectrum.

The solar interior

The Sun's photosphere is maintained at a temperature of about 6000 K by energy released inside the Sun.

Could this energy be released from burning fuel like coal or oil inside the Sun? Fossil fuels burning at a rate of 1 kg per second release energy at a rate of no more than about 50 million watts. To release energy at the same rate as the Sun (i.e. a rate of 400 million million million million watts = 4×10^{26} watts), such fuel would need to be burned at a rate of over 8 million million million kg per second (= 4×10^{26} watts/50 million watts per kg). The Sun's mass is about 2×10^{30} kg. This would be used up in about 10 000 years at a rate of 8 million million million kg per second. Clearly, the Sun's energy is not the result of burning coal or oil in its interior.

Could this energy be released from nuclear fission or from radioactive decay? These are processes in which unstable nuclei release energy on a scale which is about half a million times greater than the burning of coal or oil. On this scale, the Sun would last about 500 000 times longer than the 10 000 year estimate above. Its total lifetime would be about 5000 million years, a rather more realistic estimate than the 10 000 years if it was coal-fired.

An estimate of the age of the Sun on the basis of nuclear reactions was first made by Eddington in 1920. Instead of thinking about energy from nuclear fission where heavy nuclei are split, Eddington thought about fusing hydrogen nuclei to form heavier nuclei. He realized that the abundance of hydrogen and helium in the Sun is an important aspect of how energy is released in the Sun. He knew from experiments carried out by Francis Aston in Cambridge in 1919 that the mass of a nucleus of helium is not exactly four times the mass of a hydrogen nucleus. Aston had measured the mass of a helium nucleus and found that it was 0.8% less than the mass of four hydrogen nuclei. Aston obtained these results using an instrument called a mass spectrometer, which he invented. This device deflects a beam of ions by means of a magnetic field. The mass of each type of ion was calculated from its deflection. Eddington thought that a helium nucleus was formed as a result of forcing four hydrogen nuclei (i.e. protons) and two electrons very close together. As we shall see later, this is not quite correct, as two of the protons turn into neutrons in the process. Nevertheless, Eddington was on the right track as he deduced that the 0.8% loss of mass was because energy was released when the helium nucleus was formed from a fusion of four protons. He used Einstein's formula $E = mc^2$ to work out that a fusion rate of 1 kg per second of hydrogen would release energy at a rate of over 700 million million watts ($= 7.0 \times 10^{14}$ watts). This calculation is explained in more detail on page 105. On this scale, the Sun fuses hydrogen into helium at a rate of 0.6 million million kg per second ($= 4 \times 10^{26}$ watts$/7 \times 10^{14}$ watts per kg). The mass of the Sun is 2×10^{30} kg, enough to continue fusing hydrogen into helium for over 100 000 million years.

Eddington's remarkable estimate has since been confirmed and improved by more detailed studies. Even though Eddington was not sure how hydrogen nuclei fuse together to form helium nuclei, he had the insight to realize the significance of the mass loss that takes place when hydrogen fuses to form helium. As mentioned in Chapter 5, even now at the end of the twentieth century, scientists have still not worked out the exact mechanism by which mass is converted to energy. However, the process by which hydrogen is fused to form helium, called the proton–proton cycle, was worked out in 1938 by Hans Bethe at Cornell University. Bethe also proved that helium could be produced from hydrogen as a result of the addition of four protons one after another to a carbon-12 nucleus. The carbon nucleus first becomes a nitrogen nucleus, then an oxygen nucleus, then a heavier nitrogen nucleus before it reverts back to a carbon-12 nucleus by emitting a helium nucleus.

Heavier atoms are formed in the later stages of the life cycle of a star, when its hydrogen fuel has mostly been consumed and converted into helium. In 1953, Fred Hoyle showed that helium-4 nuclei could combine in threes to form carbon-12 at 100 million degrees. The addition of further helium-4 nuclei would produce heavier atoms including prominent elements found in stars such as oxygen-16, magnesium-24 and iron-56. Interestingly, the atomic masses of such elements tend to be multiples of four, corresponding to whole numbers of helium nuclei being fused together to form heavier nuclei.

The fusion processes, described above, are known to release energy when they occur, because the mass of the nucleus formed is always less than the combined mass of the nuclei used. The loss of mass is due to energy released when the nuclei fuse together, in accordance with Einstein's formula $E = mc^2$. Release of energy due to fusion only works for nuclei as heavy as iron. The mass of every known nucleus has been measured by mass spectrometry. For nuclei heavier than iron, the mass of the nucleus is always more than the combined mass of lighter nuclei with the same total number of protons and neutrons. Such heavy nuclei can only release energy by fission (i.e. splitting) not by fusion. Thus stars formed only from hydrogen and other light elements cannot create elements heavier than iron. Why then are heavy elements like uranium present in the Earth? This is a question we shall look at later in this chapter.

Around the core

Nuclear fusion in the Sun takes place in its core, where solar matter is under greatest pressure. Enormously high pressure is needed to force nuclei together to make them fuse. Without such pressure, two nuclei repel when they come into contact with each other because they are both positively charged. In the Sun's core, the pressure is great enough to force light nuclei so close that they fuse with each other. In the process of nuclear fusion, photons of gamma radiation are released. These photons travel away from the core, interacting with fast-moving atomic nuclei and electrons until they reach the region where nuclei and electrons are combined in the form of atoms and ions. This region forms the photosphere as its atoms and ions are continually colliding with each other as they move about. The electrons in these atoms and ions gain energy from the collisions, move up to higher energy levels as a result and then emit light photons when they fall back.

The energy produced in the core is carried mostly by gamma radiation to the photosphere. The interior of the Sun between its energy-producing core and near the photosphere is referred to as the **radiative zone** of the Sun because radiation carries the energy outwards. The matter in this zone is a gas of dense, unattached nuclei and electrons with too much kinetic energy to form atoms and ions. The inward pull of gravity on the matter in the radiative zone is counteracted by the outward pressure of this gas, provided the gas continues to be heated by a steady stream of radiation from the nuclear furnace at the core of the Sun.

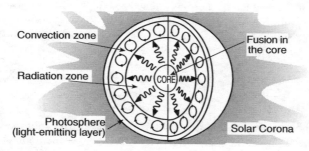

Figure 7.4   Inside the Sun

Luminosity rules

We saw on page 76 that the energy radiated per second, or luminosity of a star, is proportional to its surface area and the fourth power of its temperature. Since the surface area is proportional to the square of the radius, then the luminosity L is proportional to R^2T^4.

$$L \text{ is proportional to } R^2T^4$$

The balance between the inward pull of gravity and the outward pressure of the gas in the radiative zone can be used to show that the temperature T at the edge of the radiative zone is proportional to M/R where M is the mass of the star and R is its radius.

$$T \text{ is proportional to } M/R$$

This can be seen because temperature is a measure of the kinetic energy of a particle in a gas and M/R is a measure of the gravitational potential energy of a particle at the surface of a sphere of mass M and radius R.

Therefore, by combining the two links above,

L is proportional to M^4/R^2

Since the mass M is proportional to the volume which equals $\frac{4}{3}\pi R^3$, then we can obtain a relationship between L and M in the form

L is proportional to $M^{10/3}$

This is very close to the relationship described on page 90 between L and M that Eddington discovered since 10/3 is very close to 3. In fact, the theory can be refined to derive the rule that the luminosity is proportional to the cube of the mass. However, it should be borne in mind that stars do not fit the exact rule anyway. Nevertheless, the agreement is close enough to conclude that the radiative model of the Sun's interior is valid.

The violent Sun

The Sun is the nearest star to Earth. Fortunately, the Earth is not too near the Sun otherwise it would be baked by its heat and occasionally well and truly cooked by outbursts of solar energy in the form of solar flares and solar prominences. These are gigantic flaming outbursts that reach out hundreds of thousands of kilometres from the photosphere, sometimes lasting for hours. Solar flares cause communications problems on Earth, as huge numbers of charged particles are ejected by a solar flare. These particles spread through space and can temporarily disrupt the upper atmosphere, preventing the transmission of radio waves. Solar prominences are enormous arches of solar matter thrown up from the photosphere, lasting days or even weeks before they die down.

WARNING Never look at the Sun under any circumstances, as it will damage your eyes permanently

The life cycle of a star

The birth of a star

A star is born when clouds of hydrogen and other gases in space contract due to their own gravity. The particles in these clouds swirl round and speed up as the matter contracts and becomes more dense, the same way as a falling object speeds up as it falls. The particles collide violently with each other as the gas clouds contract, causing internal heating and a rising temperature. In effect, the gravitational energy of these clouds of matter is turned into thermal energy as the clouds contract and become denser and denser. The clouds that eventually form a star may stretch a distance of a few light years across, taking tens of millions of years to contract and heat up enough to make hydrogen nuclei fuse into helium nuclei.

A star in formation is called a **protostar**. Its core needs to reach a temperature of millions of degrees before hydrogen nuclei start to fuse to form helium nuclei. When nuclear fusion commences, the core temperature rises further as energy is released from the fusion processes in the core. The protostar becomes a fully-fledged star, continuing to release energy at its core as a result of nuclear fusion. The envelope of hydrogen and other matter round the core turns into the radiative zone, surrounded by a photosphere and a corona of hot gases. Nuclear fusion in the core maintains the core temperature and provides the necessary internal pressure to prevent continued collapse. The new star joins the Main Sequence at a point corresponding to its total mass, high up if its mass is high and low down if its mass is low. The star remains at this position on the Main Sequence for most of its lifetime.

ACTIVITY
Observing newly formed stars

The Great Nebula in Orion, M42, is an irregular cloud of glowing gas and dust heated by newly formed stars within it. It can be observed easily with low-power binoculars or a telescope almost at the centre of the lower part of Orion. The Trapezium consists of several stars spanning about 30 seconds of arc at the centre of the brightest part of the nebula. The stars of the Trapezium are thought to have formed about a million years ago from the dust and gas in the nebula.

The nebula also contains stars in formation that can only be detected from the infra-red radiation they emit.

After the Main Sequence

The energy-producing core of a star is thought to contain about 20% of the star's hydrogen. The star gradually converts the hydrogen in its core into helium, remaining at the same position on the Main Sequence until the supply of hydrogen in the core is exhausted. At this stage, the core temperature falls and the core collapses. As a result, the envelope of hydrogen surrounding the core expands to counteract the collapse of the core, changing the star into a red giant. Its change of colour is because the outer layers become cooler as they expand, so the star changes to an M or a K class star. The transition from the Main Sequence to the red giant stage is relatively rapid. **Long-period variable stars** that pulsate very slowly, such as Mira in the constellation of Cetus, are unstable red giants that are thought to have used up all the hydrogen content of their cores. Mira's brightness peaks about once every 330 days, reaching between magnitude 2 and 4 at each peak. At its faintest, Mira is invisible to the unaided eye. At a distance of over 75 parsecs from the Earth, it is actually a supergiant as its maximum diameter is over 200 times that of the Sun.

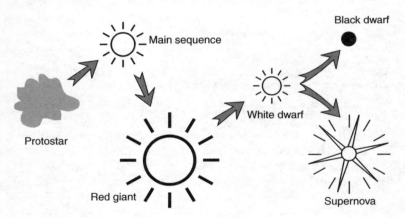

Figure 7.5   The evolution of a star

During the red giant stage of a star's existence, the contracted core is hot enough to cause helium nuclei to fuse to form heavier nuclei. When the core's supply of helium has been used up, the entire star collapses because its internal pressure drops. The star becomes much smaller than it was before its giant phase and it heats up to become white-hot as its thermal energy is much more concentrated. It is now a **white dwarf**, much smaller and much hotter than the Sun. A white dwarf is also much more dense than the Sun. The density of a white dwarf, more than 100 000 times that of water, is the result of atoms being crushed until their electrons cannot become any closer to each other.

White dwarf stars were discovered in 1910 by Henry Pickering at Harvard. He discovered that a faint tenth magnitude star, a companion of Omicron Eridani in the constellation of Eridanus the river, was spectral type A, and therefore very hot. He knew from parallax measurements that its distance is about 5 parsecs and realized its light output is less than one thousandth of the Sun's output even though it is much hotter. He was astonished when he realized that it must therefore be much smaller than the Sun, and in fact little more than twice as large as the Earth. Shortly after this first-ever discovery of a white dwarf, Sirius B was unmasked as a white dwarf. This ninth magnitude star is difficult to observe, as it is the binary companion of Sirius, the brightest star in the sky. It was discovered in 1862, well before the discovery of the first white dwarf.

Sometimes, a red giant throws off a shell of glowing gas as it collapses to form a white dwarf. Such shells of glowing gas are called **planetary nebulae**, because they were originally thought to be where planets could form. The Ring nebula M57, in the constellation of Lyra, is a ninth magnitude shell of glowing gas about half a light year in width, ejected from a much fainter dwarf star. Its ring-like appearance is because the ring is where the shell appears deepest as viewed from Earth.

Stars that are not nearing the end of their life cycles can also be unstable. A **nova** is a star that suddenly becomes much brighter then fades out gradually, appearing as a new star in the sky if it was previously too faint to be seen. Nova Aquilae appeared dramatically in 1918, within days becoming as bright as Sirius, the brightest star in the sky, and remaining visible to the unaided eye for several months. Like comet-spotting, hunting for novae is one way in which an amateur astronomer can make a name for himself or herself, as a nova is an unexpected event and professional astronomers cannot usually afford to observe the sky at random. Peter Collins, an amateur astronomer in Colorado in 1992, was the first person to spot a nova in the constellation of Cygnus the swan. Within hours of its discovery, astronomers world-wide and instruments on the International Ultraviolet Explorer were observing this nova, V1974 Cygni.

A nova is a dramatic event in which a star throws off a shell of matter, brightening by perhaps 10 magnitudes or more as a result. The expanding shell of matter is usually too faint to observe directly but its presence is evident from the broad emission lines in the star's spectrum. What causes a star to behave in this way? The Earth would be incinerated within days if the Sun went 'nova'. Don't become worried though as a nova is not a common event and is thought to be a star in a binary system. One possible model, based on data from V1974 Cygni, is where a white dwarf draws off matter from a less massive white dwarf which is its binary companion. The extra matter refuels the heavier white dwarf, causing it to overheat dramatically and, with a great outburst of light, throw off much of its accumulated matter. Some novae have been observed to give a repeat performance. In 1946, T Coronae repeated its performance of 1866 when it flared up by 7 magnitudes to reach magnitude 2.

The Sun is a main-sequence star without any companions in space. In about 5000 million years, it will swell out to become a red giant then it will collapse into a white dwarf. No more fusion will be possible and its thermal energy will radiate away gradually into space. Eventually, it will fade out and become invisible – an ignominious end as a black dwarf, like a huge cinder in space.

Supernovae

A much more dramatic fate awaits stars more massive than the Sun. Such stars end their lives with an enormous explosion, referred to as a supernova, which can outshine an entire galaxy for months. The Crab Nebula M1 in Taurus is an irregular patch of glowing gas, with filaments extending from the patch. The thin filaments appeared like the legs of a crab to Lord Rosse, who first observed them in 1844. The Crab Nebula is thought to be the result of a supernova explosion in 1054 at a distance of about 2000 parsecs from us. See Plate 2.

Supernovae are rare events. In addition to the Crab Nebula, only two other supernovae have been detected in our home galaxy, the Milky Way. One of these, known as Tycho's star, occurred in Cassiopeia in 1572 and became as bright as Venus for over a year. The other one, Kepler's star, occurred in the constellation of Ophiuchus in 1604. About 50 supernovae have been observed in other galaxies.

The most important supernova event was observed in 1987 in the Large Magellan Cloud which is an irregular companion galaxy of the Milky

Way. This supernova, referred to as SN1987A, was first observed in February 1987. It instantly became the focus of attention of astronomers in every continent who used the full armoury of scientific detectors across the electromagnetic spectrum at ground-based sites as well as on satellites. It continues to be observed, and is providing astronomers with vital information about stars. Previous records of the Large Magellan Cloud indicate that SN1987A was a blue supergiant known as Sanduleak which suddenly exploded. The supernova brightened and reached magnitude 3 after a few months. Since then, it has gradually faded as the debris of the explosion continues to disperse at enormous speed.

ACTIVITY
Observing the debris of a supernova

Use binoculars or a telescope to observe the Crab Nebula in the constellation of Taurus, about the same 'height' above the top of Orion as Orion's belt is below. This part of the sky is easily visible from November to March from the Northern hemisphere. The Crab Nebula is an intense source of X-rays and radio waves. The expanding gas cloud is about 10 light years across, with a neutron star rotating 30 times each second where the original star was.

Why do massive stars come to such a dramatic end?

In 1930 this question was considered in depth by a young Indian student, Subrahmanyan Chandrasekhar, as he journeyed by ship from Bombay to London in order to take up a fellowship at Trinity College, Cambridge. Chandrasekhar combined the relativistic theory of the electron with the theory of white dwarfs and showed that a white dwarf would collapse if its mass was more than 1.44 solar masses, which became known as the **Chandrasekhar limit**. Any white dwarf below this limit would fade out gradually to become a black dwarf. Above this limit, the star would collapse. Eddington challenged the theory on the grounds that a complete collapse to a point is a physical impossibility, and that no known process could stop the collapse. In fact, within two years, an important discovery in nuclear physics was to provide the key to overcoming Eddington's objection and to justify Chandrasekhar's theory without reservation. We will meet this key discovery after an update about the nucleus next.

About the neutron

Discoveries in science often have unforeseen consequences, sometimes beyond the realms of our imagination. In the first decades of the twentieth century, Lord Rutherford discovered that most of the mass of an atom is concentrated in a positively charged nucleus. He concluded that the nucleus of hydrogen, the lightest atom, is a single particle which became known as the proton. He and his colleagues worked out that the nucleus of helium has a mass which is four times that of hydrogen and a charge that is double the charge of the hydrogen nucleus. From this comparison and other similar comparisons, he predicted the existence of the neutron, a neutral particle of about the same mass as the proton. His prediction remained no more than that until 1932, when one of his former students, James Chadwick, discovered that a beam of protons knocks neutral particles of about the same mass from a wax target. Chadwick had discovered the neutron predicted by his former boss. The neutron became the subject of intense experimental research and in 1938 Otto Hahn and Fritz Strassman in Berlin discovered the principle of induced fission by neutrons. Within a decade of Chadwick's discovery, the first nuclear reactor in 1942 had been constructed and successfully tested by Enrico Fermi in Chicago. Just three years later, in 1945, the first atomic bomb changed the course of history.

What causes a supernova?

Walter Baade and Fritz Zwicky realized the significance of Chadwick's discovery for astronomy and in 1934 they published a paper in which they advanced the idea of a neutron star. They knew that a supernova is a much rarer and much more energetic event than a nova, and they put forward the theory that a supernova is an event that happens when an ordinary star changes into a **neutron star**. The density of such a star is far greater than the density of a white dwarf. Calculations subsequently have shown that a neutron star of mass greater than that of the Sun has a diameter of little more than 10 kilometres. The strength of gravity at the surface of a neutron star is so great that light is bent by it, almost strong enough to prevent light escaping. Theoreticians at the time knew that Einstein's general theory of relativity predicted the existence of black holes, objects so massive that not even light could escape. Clearly, the discovery of the neutron and the prediction of neutron stars turned the black hole from a mathematical prediction into a physical possibility. We shall return to black holes in more detail in Chapter 9.

Zwicky began a prolonged sky search for supernovae using a wide-angle 18-inch telescope he constructed at Mount Palomar in California. Within a few years, he had discovered more than ten supernovae in nearby galaxies. Enough observational data had been gathered within a decade for astronomers to categorize supernovae into two types. Type I supernovae were found to be identical in terms of their spectra and brightness variation. Other supernovae were classed as type II and were found to be due to the collapse of stars of mass in excess of about 8 solar masses. We shall see in Chapter 10 that a type I supernova in a distant galaxy is like a 'milestone' that can tell us how far away the galaxy is.

Astronomers reckon that type I supernovae occur when a white dwarf attracts mass from a binary companion, taking it over the Chandrasekhar limit of 1.4 solar masses. The white dwarf then suddenly collapses into a neutron star as the electrons and protons in the white dwarf are forced to react, and become neutrons in a process like a reverse beta decay. The sudden collapse at the centre of the white dwarf causes a shock wave to spread out through the outer layers, which are blasted off in a cataclysmic explosion, which is observed as the supernova. The inner core continues to collapse, leaving a neutron star as a remnant containing a fraction of the original mass, with a density of the order of a million million times greater than that of the Earth.

What experimental evidence exists for neutron stars? The theory that a neutron star is a remnant of a supernova provides an explanation of supernova events but does not amount to observational evidence. In 1967, Jocelyn Bell, a Cambridge research student, discovered a source of regular radio pulses in the sky. Within a year, twenty more similar stars, referred to as **pulsars**, were discovered. Astronomers worked out that a fast-spinning neutron star emits a radio beam that sweeps round like a lighthouse beam as the star spins. Each time the beam sweeps past the Earth, a pulse of radio waves is detected from the star. The neutron star at the centre of the Crab Nebula is a pulsar. We will meet pulsars again in Chapter 9.

An elementary puzzle

Nuclear fusion inside stars builds up elements only as heavy as iron. This is because energy is needed to form elements heavier than iron. When light nuclei fuse together, energy is released to keep the temperature in the star hot enough for the fusion process to continue. Nuclei heavier than iron do not form from light nuclei inside stars like the Sun because energy is needed for such processes. How then are such nuclei formed? Put in simple terms, how were elements such as gold, lead and uranium formed? Nuclear fusion in a star does not provide the answer. The formation of elements heavier than iron was a puzzle for nuclear physicists and astronomers several decades ago. Supernovae provided the answer as it was realized that the addition of neutrons to light nuclei in a supernova explosion could create nuclei as heavy as uranium, since many of the added neutrons would decay into protons. Extensive calculations, using computers, were carried out to show that such processes could indeed account for the abundance of each element in nature. The Sun and its solar system must therefore have formed from the debris of a previous star that exploded as a supernova over 5000 million years ago.

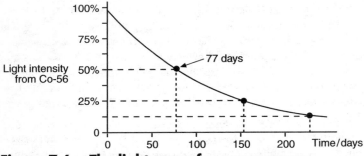

Figure 7.6   The light curve for a supernova

Theoretical calculations alone do not add up to conclusive proof. This was obtained from supernova SN1987A as a result of measuring the decrease of light intensity of its prominent spectral lines. Each line is due to light emitted by a known element, so the decrease of abundance of that element is monitored from the light intensity of a relevant line. In particular, the light intensity from cobalt-56 was found to decrease by 50% every 77 days. This agrees with the radioactive half-life of cobalt-56 which is known from laboratory measurements to be 77 days. Measurements on other elements present in SN1987A give similar agreements. Clearly, a supernova is a chemical cradle. The Sun will not end its life cycle as a supernova but will fade out to become a black dwarf. We shall return to the question of whether or not supernovae were more prominent in the early Universe in Chapter 12.

Star lifetimes

How old are the stars? Are the most luminous stars the oldest? How much longer will the Sun continue to be on the Main Sequence? We saw on page 93 that if all the Sun's hydrogen were converted to helium, the Sun would continue to shine for 100 000 million years. This estimate is based on the fact that the mass of the helium nucleus is 0.8% less than the mass of four hydrogen nuclei (i.e. protons). Let's look at the main steps in this estimate in detail.

- **Step 1**. The fusion of a mass of 1 kilogram of hydrogen causes a mass loss of 8×10^{-3} kg. Using Einstein's mass energy formula, the energy released per kilogram of hydrogen is therefore 7×10^{14} J ($= mc^2 = 8 \times 10^{-3}$ kg $\times (3 \times 10^8$ m/s$)^2$, where the speed of light, $c = 3 \times 10^8$ m/s).
- **Step 2**. The mass of the Sun is 2×10^{30} kg so the total energy it could release is 14×10^{44} J ($= 7 \times 10^{14}$ W $\times 2 \times 10^{30}$ kg).
- **Step 3**. Since the Sun is known to radiate energy at 4×10^{26} watts, it could last for about 3.5×10^{18} seconds ($= 14 \times 10^{44}$ J$/4 \times 10^{26}$ W) which is about 100 000 million years.

Chandrasekhar proved that a Main Sequence star like the Sun would convert about 12% of its total hydrogen into helium before it leaves the Main Sequence to become a red giant. Since most of its life cycle is spent on the Main Sequence, the lifetime of the Sun is therefore 12 000 million years. Its present age is known to be about 5000 million years from radioactivity measurements so the Sun is about half-way through its life cycle – a middle-aged star with an unspectacular future!

The lifetime of a star depends on its mass and on its luminosity or absolute magnitude. The lifetime of a star that is ten times as luminous as the Sun and has an equal mass will last one tenth as long as the Sun, because it uses up the same amount of fuel at ten times the rate. A star that is one thousand times as luminous as the Sun and has a mass ten times greater than the Sun will only last 1/100th as long as the Sun. Comparing the lifetimes of different stars is like comparing the range of vehicles in terms of fuel consumption and fuel capacity. For example, a luxury car which uses fuel at three times the rate of a family saloon has the same range if its fuel capacity is also three times bigger.

■ Near the top of the Main Sequence, the O and B class stars are about 10 magnitudes more luminous than the Sun and about 50 times more massive. Since 10 magnitudes corresponds to a factor of 10 000 in the light output, the lifetime of these stars is 1/200th of that of the Sun, so they will use up their fuel within about 50 million years. These brilliant blue stars will be white dwarfs or even neutron stars long before the Sun runs out of hydrogen.

■ Near the foot of the Main Sequence, the stars are much less luminous than the Sun and less massive. An M-class Main Sequence star is about 5 magnitudes less luminous than the Sun so it radiates energy at about 1/100th of the Sun's rate. Its mass is about one tenth of the mass of the Sun. Therefore, although it has only one tenth the amount of fuel the Sun has, it will last ten times longer than the Sun, because it uses its fuel at one hundredth the rate the Sun does. M-class Main Sequence stars, feeble though they are compared with the Sun, will still be shining long after the Sun has ceased to exist.

Knowledge of the lifetime of an M-class Main Sequence star does not mean we can say how old the star is. It could have formed recently and have many thousands of millions of years to go on the Main Sequence. Alternatively, it could have formed many thousands of millions of years ago and be near the end of its life on the Main Sequence. We will return to the question of the age of a star in Chapter 10 where we look at the evidence that the oldest stars are almost as old as the Universe. In fact, for a time, the estimated age of these stars was greater than the estimated age of the Universe!

Summary

■ A **binary** is two or more stars in orbit about each other.
■ A **spectroscopic binary** cannot be resolved into two stars by observation. The lines of its spectrum repeatedly open and close.
■ **Kepler's 3rd Law applied to binaries:**
    Total mass $\times$ (Time period)^2 = (Separation)^3
■ **Luminosity** is the energy per second radiated by a star. The luminosity of a Main Sequence star is approximately proportional to the cube of its mass.

■ **The energy released inside a star** is due to the fusion of hydrogen nuclei into helium.

■ **The age of a star** can be estimated from its luminosity and its mass. A Main Sequence star converts about 10% of its hydrogen into helium. The more massive a Main Sequence star is, the shorter its life is. The Sun is about 5000 million years old.

■ **The evolution of a star**

A protostar is a star in formation.

A new star moves onto the Main Sequence where it remains for most of its life.

The greater the mass of a star, the higher its position on the Main Sequence.

A red giant is a helium-burning star which is formed when the core of a Main Sequence star runs out of hydrogen.

A white dwarf is formed when a red giant runs out of helium.

A supernova is when a white dwarf collapses. This happens if the mass of the white dwarf exceeds 1.4 solar masses. Heavy elements are formed in a supernova explosion.

8 | TOWARDS THE EDGE OF THE UNIVERSE

So far we have looked at how astronomers have used the parallax method to measure the distance to stars as far away as 100 parsecs or so. On the scale of our solar system, this distance is enormous at over 20 million astronomical units. Yet a distance of 100 parsecs on the scale of the Milky Way galaxy is about the same as a grain of sand to a large sandcastle, and on the scale of the Universe about the same as a grain of sand to a large desert! In this chapter, we shall look at how astronomers have measured the distances to objects billions of light years away. We will also look at how astronomers deduced that the Milky Way is just one of countless galaxies in the Universe. We shall see how variable stars, nova and supernova, star clusters and galaxies themselves have all been used as 'standard candles' to determine distances to the edge of the Universe. Most importantly, we meet the work of Edwin Hubble who deduced from his observations the law that the distant galaxies are moving away from our galaxy at speeds in proportion to their distances. This and other discoveries, made from observations in the early decades of the twentieth century, led to controversy and debate about the origin of the Universe itself. Observations and measurements repeated and refined using improved equipment led to greater confidence in the conclusion drawn from Hubble's law that the Universe is expanding. We will also look at the discovery of quasars and why they seem to be among the most distant objects ever observed, perhaps the result of galaxies in the early Universe in collision.

Mapping the Milky Way

On a clear dark night away from street lighting, the Milky Way is seen as a faint luminous irregular zone across the sky, its width, colour and intensity varying in position although not in time. We see it now much as it was seen by Ptolemy in about AD 150 in one of the earliest known descriptions in which he referred to its name being because it is as 'white as milk'. Many centuries later, Galileo was the first person to observe it using a telescope, discovering that it contained a multitude of stars too faint for the unaided eye to see. Its appearance as an irregular zone across the sky was explained by William Herschel about two centuries ago, who correctly deduced its disc-like shape although he was unable to estimate its actual width. We know that the Milky Way is about 40 000 parsecs in diameter, and that the Sun is about two-thirds of the distance from the centre to the edge of the galaxy in the galactic disc. Also, we know that the galaxy includes a halo of stars and clusters of stars above and below the disc. Let's look at how this picture has been built up.

The centre of the Milky Way lies beyond the dust clouds of Sagittarius. Light cannot penetrate these clouds but radio waves can. Hydrogen gas in the disc of the Milky Way emits radio waves. Using radio telescopes, astronomers have been able to piece together a map showing the distribution of hydrogen in the galaxy. The map confirms the picture of the Milky Way built up by optical astronomers. Over the period from 1906 to about 1920, Jacob Kapteyn surveyed the stars in selected areas of the sky and eventually concluded that the disc of the Milky Way is about 12 000 parsecs in diameter, much less than present estimates, and about 1500 parsecs thick. To arrive at this conclusion, he used the information on stars at known distances to work out that the absolute magnitude of most stars is between about 5 and 10. Then he applied this range to all the other stars in his survey and deduced their distances and distribution.

Measuring distances beyond the range of the parallax method

Distance measurements to stars beyond the range of the parallax method were made using various methods, including using the spectrum of a star to decide where on the Hertzsprung–Russell (HR) diagram it ought to be located. A star that can be classified from its spectrum can then be placed on the HR diagram and its absolute magnitude M could then be determined from its position on the diagram. Its **distance modulus** (m – M) could be used to determine its distance. For example, suppose a star of magnitude 7 is a K-class red giant. Its absolute magnitude is therefore 0 (using the HR diagram on page 74) so its distance modulus is 7 (= 7 – 0). Using Figure 6.2, its distance is therefore 250 parsecs.

The distance to a cluster of stars can be determined more accurately than the distance to an individual star. The stars in the cluster are plotted on an HR diagram, using the observed magnitude instead of the absolute magnitude of each star. Provided there are sufficient stars, their locations on the diagram form a main sequence pattern which is displaced vertically on the diagram from the correct position of the Main Sequence. The distance modulus (m − M) is the vertical displacement between the correct position of the Main Sequence and the cluster's main sequence. Using these methods, the distances to stars as far away as 10 000 parsecs have been determined. Beyond this distance, the spectra of stars are too faint to record and clusters are too rare. A different approach is needed. This was provided in 1912 by Henrietta Leavitt using variable stars whose brightness varies in a regular manner, stars known as **Cepheid variables**.

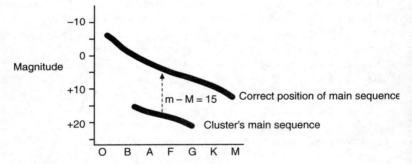

Figure 8.1(a)   Main sequence fitting

Cepheid variables

In addition to estimating the size of the Milky Way, astronomers also wanted to know if nebulae such as the Andromeda nebula and the Magellan Clouds are within or outside the Milky Way. Henrietta Leavitt, at Harvard, observed over 1700 variable stars in the clouds and identified over two dozen as **Cepheid variables**. This type of star was discovered over a century earlier by John Goodricke, who found that the brightness of a star in the constellation of Cepheus, δ (pronounced delta) Cephei, varied between magnitude 3.5 and 4.4 regularly with a period of 5.4 days. Other stars with brightness variations over similar periods were observed in different parts of the sky, all referred to as cepheid variables. The Pole Star is a cepheid variable with a period of 4 days and a 0.1 magnitude variation – too small to detect visually.

Henrietta Leavitt systematically measured the light curves of 25 Cepheid variables in the Small Magellan Cloud. A light curve of a star is a graph which shows how the magnitude of the star changes with time. Leavitt discovered that the longer the period of a Cepheid variable, the greater its mean magnitude. By plotting the mean magnitude on the vertical scale of a graph and the period on the horizontal scale, she discovered that the mean magnitude increased smoothly as the period increased.

Leavitt's results, published in 1912, provided astronomers with the key to measuring distances far beyond the Milky Way. The significance of these results at that time lay in the fact that all the Cepheid variables in the Small Magellan Cloud are at the same distance from us – just as everyone in a city thousands of kilometres away is virtually at the same distance from you, give or take a few kilometres. If the distance to the Small Magellan Cloud had been known by Henrietta Leavitt, she could have converted the observed magnitudes to absolute magnitudes. This conversion could then have been used to find the distance to any other Cepheid variable simply by measuring its period and using Leavitt's results to determine its absolute magnitude. Its distance could then be calculated from this and its mean observed magnitude, as explained on page 71. Nevertheless, Henrietta Leavitt established that there is a well-defined link between the period and the mean absolute magnitude of a Cepheid variable, and that the link could be used to measure distances far in excess of parallax measurements.

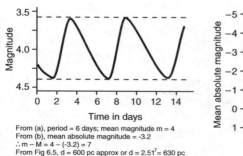

From (a), period = 6 days; mean magnitude m = 4
From (b), mean absolute magnitude = -3.2
∴ m − M = 4 − (-3.2) = 7
From Fig 6.5, d = 600 pc approx or $d = 2.51^7 = 630$ pc

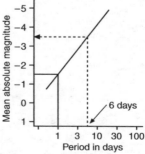

(i) Light curve for a Cepheid variable; (ii) Calibration graph
Figure 8.1(b)   Cepheid variables

One method of finding the distance to the Small Magellan Cloud was to measure the distance to one or more Cepheid variables at known distances in the Milky Way. This distance could then be used to calculate absolute magnitudes and then to plot a graph of absolute magnitude against period. This graph could then be used to give the absolute magnitude of any Cepheid variable in the Small Magellan Cloud (or anywhere else) from its period. By comparing the observed magnitude with the absolute magnitude, the distance to the Cloud could then be found. This method was adopted by Hertzsprung in 1913 who obtained a result of 12 000 parsecs for the distance to the Small Magellan Cloud. Subsequent measurements using a different calibration method now put the distance at over 60 000 parsecs.

Globular clusters

The discovery of the link between absolute magnitude and period for Cepheid variables offered astronomers a method of measuring the size and structure of the Milky Way. In 1918, Harlow Shapley used the method to measure the distance to a large number of globular clusters. A globular cluster is a tight spherical cluster of millions of stars held together by their own gravity. See Plate 4. These clusters are observed above and below the plane of the Milky Way, mostly in the approximate direction of the constellation Sagittarius. Although the stars in a globular cluster appear very close together, the diameter of a cluster is of the order of 30 parsecs or about 100 light years. The average spacing of the stars in a cluster is of the order of a light year, a little less than the distance between the Sun and its nearest neighbour, Proxima Centauri. The sky on a planet at the centre of the cluster would appear filled with stars in every direction.

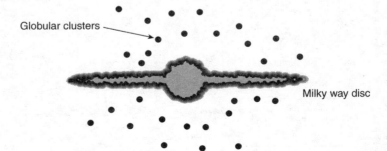

Figure 8.1(c)   Globular clusters

Clusters can be resolved into individual stars to identify variable stars which can then be timed and used to find the distance to each cluster. Shapley found that the clusters are distributed about a centre, in the direction of Sagittarius, which he assumed to be the galactic centre. He worked out the distance to the centre at more than 15 000 parsecs. Subsequent measurements taking account of the absorption of light by dust in the Milky Way have shown the distance to be about 5000 parsecs too large.

Shapley also estimated the diameter of the Milky Way at about 90 000 parsecs, about ten times more than Kapteyn's earlier estimate. The conflict between the two estimates deepened as a result of observational evidence from novae. Astronomers had discovered that novae in the Andromeda nebula M31 were much fainter than novae in the Milky Way. Their calculations, based on observed differences of 10 magnitudes, indicated that M31 is perhaps 300 000 parsecs or about a million light years away, well outside the Milky Way. The 1885 nova in Andromeda, however, was much much brighter than the others observed in Andromeda, perhaps indicating that Andromeda is actually inside our galaxy. The problem was settled by Edwin Hubble in 1925, who observed and measured 12 Cepheid variables in M31. He compared his results with Henrietta Leavitt's results on the Small Magellan Cloud and proved conclusively that M31 lies about 500 000 parsecs from us, well beyond the Milky Way. Estimates based on novae were subsequently carried out again in more detail and its distance is now reckoned to be over 650 000 parsecs which is more than two million light years.

ACTIVITY
Observing neighbouring galaxies
1 The Andromeda galaxy M31

If you are in the Northern hemisphere, use binoculars or a telescope to observe M31 which lies about 15 degrees from Cassiopeia in the opposite direction to the Pole Star. Its bright central part is surrounded by fainter coloured spiral arms, over 150 000 light years across. See Figure 2.2 and Plate 5.

2 The Magellan Clouds

The clouds are about 15 degrees from the South Celestial Pole and therefore can only be seen from the Southern hemisphere. The two clouds form a right angle with the South Celestial Pole. The more intense regions in both clouds are the tell-tale signs of new stars in formation. The most recent supernova, SN1987A, lies in the Tarantula nebula, which is part of the Large Magellan Cloud.

The structure of the Milky Way galaxy

The photographic mosaic of the Milky Way in Plate 3 reveals a thin, irregular belt of dust clouds and filaments against a band of glowing gas and stars stretching round the celestial sphere. The stars are more concentrated in the direction of the constellation of Sagittarius and least in the opposite direction which is towards the constellations of Auriga and Perseus. These observations are enough to deduce that the Milky Way contains a disc of dust, glowing gas and stars which the Sun is in. Because the concentration of stars is greatest in the direction of Sagittarius, we are positioned away from the centre. The 'thickness' of the disc increases from about 1000 parsecs to a bulge more than twice as wide at the galactic centre, indicating that there is an enormous concentration of stars, dust and gas at the galactic centre. Relatively few individual stars lie above or below the disc of the Milky Way, although over 100 globular clusters each containing millions of stars lie above and below the galactic plane.

The methods of measuring the distances to many stars in the Milky Way and to its globular clusters were described on pages 109–110. The globular clusters were found to be centred on a region in the constellation of Sagittarius. Assuming the clusters are distributed more or less uniformly about the galactic centre, the centre of the globular clusters is the centre of the galaxy. Even though the galactic centre cannot be seen directly, because it is obscured by the dust in the galactic plane, its position was worked out from the distances to the globular clusters and their distribution.

A pattern also emerged in terms of the type of stars in different directions as astronomers observed that open clusters lie in the galactic plane away from the galactic centre and contain many hot blue stars rich in heavy metallic elements. In comparison, such stars are absent in globular clusters which lie well above or below the galactic plane. These clusters include many red giants without metallic elements present. Metal-rich red giants are, however, present in the central parts of the galaxy.

The hot blue stars were referred to as **population I stars** and the red giants in the globular clusters as **population II stars**. Since hot blue stars on the Main Sequence are short-lived compared to red giants, population I stars are young stars whereas population II stars are much older. The reason for the abundance of metal in the hot young stars is because these stars have formed from the debris of previous generations of short-lived massive stars.

The same pattern is evident when we look at other galaxies such as the Andromeda galaxy, M31 or the Whirlpool Galaxy, M51. These galaxies possess spiral arms which, when photographed in blue light, are traced out by many population I stars in the arms. In contrast, many population II stars are evident in globular clusters and the central disc of the galaxy when the galaxy is photographed in red light.

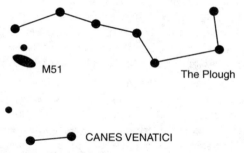

Figure 8.2(a)   Finding M51

Why should young metal-rich stars lie in the galactic disc away from the centre whereas old metal-deficient stars are concentrated in globular clusters and at the galactic centre? One possible explanation is that population II stars formed from a vast cloud of hydrogen gas of galactic proportions when the Universe was no more than a billion or so years old. These stars formed globular clusters over the same time span as a rotating galactic disc formed from the rest of the cloud. The more massive population II stars formed in the disc eventually exploded as supernovae to scatter material including heavy elements into space. Population I stars continue to form in the disc from supernovae debris as well as from hydrogen in the disc.

Does the Milky Way have spiral arms like M31 and M51? Many galaxies do not possess spiral arms and we cannot see in photographs like Plate 3 direct evidence for spiral arms. A photograph of the disc from a space probe above the disc would settle the question. However, the disc is of the order of 1000 parsecs thick so the space probe would need to travel a distance of more than 3000 light years to reach a position above or below the disc. Clearly, such a vast distance, albeit small on the galactic scale, is well beyond the range of any space probe. Nevertheless, astronomers have worked out that the Milky Way does possess spiral arms. We will look at the evidence for this picture later in this chapter.

Where is the Sun taking us?

The Sun is moving at a speed of over 20 kilometres per second through space relative to neighbouring stars. The entire solar system is moving with the Sun through space. Each day, the Sun takes us a distance of more than 1.7 million kilometres through space. In one year, we cover a distance of about 4.2 AU because of the Sun's motion through space. The nearest star lies almost seventy thousand times this distance. You will no doubt be relieved to know that a collision between the Solar System and another star is very remote – assuming there are no black dwarfs in the Sun's path!

The Sun's speed and direction relative to its neighbours may be calculated by measuring the speed and direction of as many neighbouring stars as possible. After correcting for the Earth's motion round the Sun, the star speeds and directions are all relative to the Sun. The **average** speed and direction of all these stars was then assumed to be on account of the Sun moving in the opposite direction. The same principle is evident when you walk outdoors when it is raining. Even if the rain is falling vertically, you still need to tilt an umbrella in the direction in which you are moving otherwise you will soon get wet. The result of such an exercise gives the Sun's speed as 20 km/s in a direction towards the constellation of Hercules. The Sun therefore moves a distance of 6.3×10^{11} metres or 4.2 AU each year through space.

The cosmic year

The shape of the Milky Way galaxy suggests that the galactic disc is rotating about the centre of the galaxy. The speed and direction of motion of stars in different parts of the galactic disc indicate the stars in the disc are rotating about the hub of stars in the galactic centre. The Sun's speed and direction, measured relative to the globular clusters of the Milky Way, instead of relative to neighbouring stars in the galactic disc, tell us that the Sun and its neighbours are moving round the galactic centre at a speed of 220 km/s which is about 46 AU per year. **One cosmic year**, defined as one complete trip round the galactic centre by the Sun, would take about 240 million years since the Sun is about 8500 parsecs away from the galactic centre. The circumference of its 'orbit' is about 53 000 parsecs or 11 000 million AU which would take about 240 million years to cover at a speed of 46 AU per year.

How many stars are in the Milky Way?

The time taken for one galactic orbit can be used to estimate how many stars are present in the hub of the Milky Way galaxy. To do this, we use Kepler's 3rd Law from page 89 in the following form,

$$\textbf{Mass} \times (\textbf{Time period})^2 = (\textbf{Radius of orbit})^3$$
(in solar masses)   (in years)   (in AU)

Using a value of 1750 million AU (= 8500 parsecs) for the radius of orbit, and a value of 250 million years for the time period, estimate the total mass in solar masses. Assuming the average mass of a star is one solar mass, you should find that the hub of the Milky Way contains almost 100 000 million stars. If you could see these stars individually (which we can't as they are obscured by galactic dust) and took one second to look at each one, your task would take almost 3000 years!

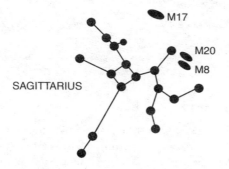

Figure 8.2(b)   Galactic nebulae in Sagittarius

ACTIVITY

Searching for spiral arms

Does the Milky Way possess spiral arms? We know the galactic disc is rotating and we know that there are many hot young stars in the galactic disc. The presence of nebulae containing hot young stars and glowing gas clouds is a hint that the galactic disc possesses spiral arms. Charles Messier, whom we met on page 9, catalogued several nebulae of this type. If the Milky Way galaxy has spiral arms, it is reasonable to expect galactic nebulae to be in the spiral arms. Such nebulae are perhaps distance indicators to the spiral arms. See if you can spot these nebulae using binoculars or a telescope.

The Lagoon Nebula, M8, in Sagittarius, is a diffuse mass of glowing gas and dust more than 30 parsecs across at a distance of more than 1500 parsecs in the direction of the galactic centre. The reason for its name is the dark ring that almost encircles a glowing gas cloud in the centre.

The Swan Nebula, M17, also in Sagittarius and less than 10 degrees from M8, lies about 1800 parsecs from the Sun. Its resemblance to a swan on water is perhaps more evident to anyone who has witnessed a fast-approaching swan!

The Trifid Nebula, M20, in Sagittarius too, is a more distant fainter neighbour of M8, remarkable for three dark bands like spokes joined at the centre of the nebula.

The Orion Nebula, M42, is easily visible with binoculars or a low-power telescope. We met M42 on page 19 in the section on star formation. It lies of the order of 500 parsecs from the Sun.

Invisible astronomy

The Earth's atmosphere shields us from gamma radiation and other forms of radiation from the Sun. Nature permits us to see the stars from the ground because the atmosphere is transparent to light. In fact, the atmosphere also transmits radio waves without significant absorption, which is why large radio telescopes on the ground can detect radio waves from space. Other forms of electromagnetic radiation from space can only be detected by means of suitable instruments mounted on space probes or on satellites or rockets.

Radio waves from space were first detected in 1933 by Karl Jansky, an American radio engineer, who was employed by Bell Telephone Laboratories. Working on a project to identify natural radio sources, Jansky detected radio waves at 20.5 MHz from the direction of Sagittarius. In 1940, Grote Reber, another American radio engineer who was also an amateur astronomer, detected radio waves at 160 MHz coming from along the galactic disc. In 1942, James Hey and his colleagues working on the development of radar in Britain found that the Sun was a strong source of radio waves. Four years later, the same group discovered a strong radio source in the constellation of Cygnus which became known as Cygnus A. Further radio sources were soon discovered, including radio galaxies and supernovae that are strong radio sources. Investment in better radio telescopes and improved detection methods led to the unexpected discovery of quasars and pulsars, which are radio sources of entirely different types to each other and to previous known radio sources. We will meet quasars again later in this chapter and return to pulsars in Chapter 9, when we look at black holes.

What causes radio waves from space? Radio waves were unknown until 1885 when the German physicist, Heinrich Hertz, showed how to produce and detect radio waves. He measured the speed of radio waves in air and found it is the same as the speed of light in air. By 1910, radio waves were being used to transmit information across the globe. A transmitter aerial transmits radio waves when electrons are forced to oscillate up and down the aerial. A receiver aerial detects radio waves as a result of electrons in the aerial being forced by the radio waves to oscillate up and down the aerial, creating a small alternating voltage in the aerial.

Synchrotron radiation is electromagnetic radiation emitted by a charged particle that is forced by an intense magnetic field to move round in circles. Radio waves from discrete sources in space are due to very strong magnetic fields causing hydrogen ions and other charged particles to whirl round in circles and radiate synchrotron radiation over a very wide band of radio frequencies.

Galactic radio waves from the Milky Way are particularly intense at a wavelength of 21 cm which corresponds to a frequency of about

1429 MHz. This is because the galactic disc contains enormous quantities of hydrogen gas as well as dust. The electron in a hydrogen atom either aligns its 'spin' in the same direction as the spin of the nucleus or in the opposite direction which requires less energy. The higher energy state is unstable and is reached as a result of inter-atomic collisions. When an atom in a 'parallel spin' state reverts to a stable state, it emits a burst of radio waves at 1429 MHz, corresponding to the exact energy difference between the two spin states of the hydrogen atom. By measuring the intensity of 21 cm radio waves at intervals along the galactic disc, the amount of hydrogen in each direction can be found, because radio waves pass through the dust clouds unlike light which is absorbed.

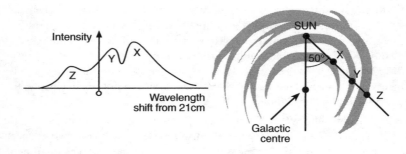

(i) Intensity profile at 50°   **(ii) 21 cm radio map**

Figure 8.3   Using radio waves to map the Milky Way

More significantly, since the galactic disc is rotating, the radio waves from hydrogen atoms in the galactic disc are doppler-shifted. If the disc did not consist of spiral arms, the radio waves would be doppler-shifted to a different wavelength according to the direction. Some of the results when the measurements were first made in 1958 are shown in Figure 8.3. The presence of three peaks at 50° either side of the galactic centre indicates three spiral branches including the branch containing the Sun. A more detailed analysis of all the measurements enables the hydrogen gas in the Milky Way disc to be mapped, to show the presence of spiral arms very clearly.

Plate 1 **Comet Hale Bopp**
(Frank Zullo/Science Photo Library)

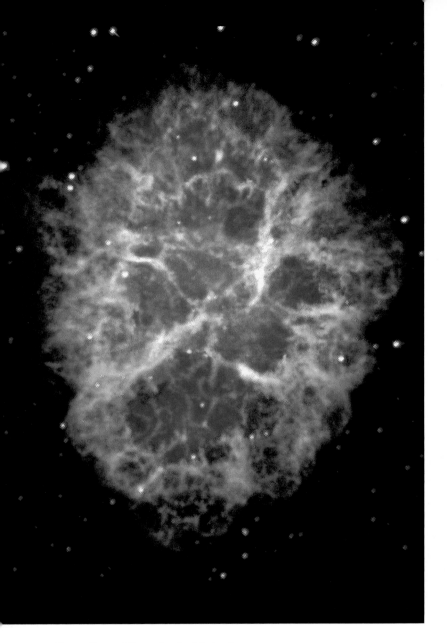

Plate 2   The Crab Nebula
(Jeff Hester and Paul Scowen, Arizona State University/Science Photo Library)

Plate 3   The Milky Way
(Dr Fred Espenak/Science Photo Library)

Plate 4   M3, a globular cluster
(NOAO/Science Photo Library)

Plate 5   The Andromeda Galaxy, M31
(Tony Hallas/Science Photo Library)

Plate 6   Gravitational lensing
(Space Telescope Science Insitute/NASA/Science Photo Library)

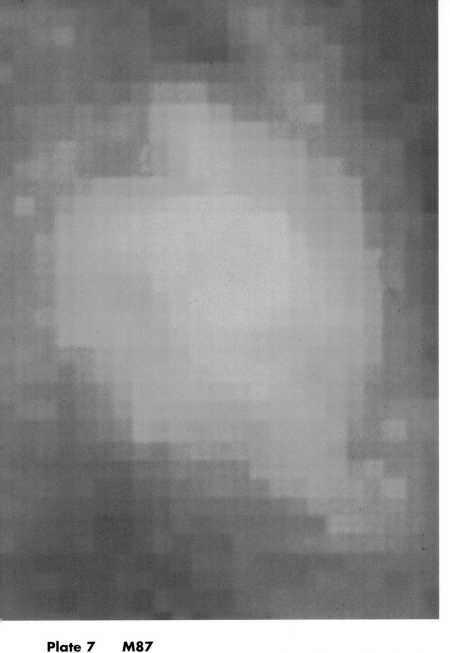

Plate 7 **M87**
(Space Telescope Science Institute/NASA/Science Photo Library)

Plate 8   M82, a galactic survivor
(Chris Butler/Science Photo Library)

Plate 9   The Large Hadron Collider
(David Parker and Julian Baum/Science Photo Library)

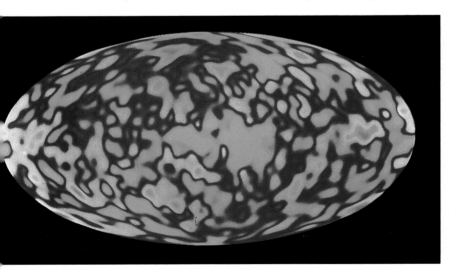

Plate 10   COBE's microwave map of the Universe
(NASA/Science Photo Library)

The Sun lies in one of the spiral arms of the Milky Way galaxy referred to as the Orion arm. When you look in the direction of the Milky Way near the constellation of Orion, you are looking along the Orion arm. The next inner spiral arm is called the Sagittarius arm because it lies in the direction of the constellation of Sagittarius. The nebulae in the constellation of Sagittarius lie over 1500 parsecs away in the Sagittarius arm. Light takes 3.26 years to travel a distance of 1 parsec, so when we look at the nebulae in Sagittarius, we are observing them as they were thousands of years ago. The next outer spiral arm to the Orion arm is called the Perseus arm because it lies in the direction of the constellation of Perseus. Stars are present in the gaps between the spiral arms, although less concentrated than in the spiral arms which are essentially defined by dust, gas and hot young stars.

Beyond the Milky Way

Clusters and clusters

In the previous section we looked at how astronomers in the first half of the twentieth century were able to prove that nebulae such as the Magellan Clouds and the Andromeda Nebula M31 lie beyond the Milky Way galaxy. Red giants, novae, supernovae and Cepheid variables have all been used to estimate these distances since the absolute magnitude of these light sources are known from identical light sources at known distances in our own galaxy. Many of the discoveries and deductions about other galaxies were the result of observations carried out using a small number of very powerful optical telescopes, which enabled astronomers to observe fainter and fainter objects in the sky. For example, the 250 cm reflector telescope at the Hale Observatory in California remained the world's largest telescope for 30 years after its completion in 1918. At over 300 times wider than the eye pupil, it is capable of collecting 90 000 times as much light as the unaided eye and therefore enables 18th magnitude objects to be seen directly. Using photographic techniques instead of direct observations, even fainter objects can be detected. We will leave the details of how a telescope works until Chapter 10 when we look at the impact of the Hubble Space Telescope.

The Milky Way galaxy and Andromeda are members of the **local cluster** of galaxies. This group of about 25 galaxies forms a cluster of the order of about one million parsecs across. Andromeda M31, the largest member of the cluster, is about 700 000 parsecs from the Milky Way galaxy. On a scale where one millimetre represented twenty parsecs, the local cluster would fit into a very large sports hall, with Andromeda and the Milky Way the size of tables at opposite ends. The Sun would be less than the size of an atom on this scale! The other galaxies in the local group include the Triangulum Spiral M33 and several irregular galaxies, including the Magellan Clouds. The Triangulum galaxy M33 is also referred to as the Pinwheel galaxy because of its resemblance to a snapshot of a pinwheel firework. It lies about ten degrees from M31 in the small constellation of Triangulum, named because the three most prominent stars of the constellation form a small, skinny triangle. M33 lies near the tip of the triangle a little further away than its near-neighbour galaxy M31. Seen from M33, the Andromeda galaxy would stretch almost 30 degrees across the heavens! The local group also includes a number of dwarf galaxies which are much smaller and fainter than the Milky Way galaxy. These dwarf galaxies are elliptical in shape and contain a large proportion of red giant stars. They show no evidence of spiral arms, and probably constitute less than two or three per cent of the total mass of the local group. The Fornax galaxy, in the constellation of Fornax, is a dwarf elliptical galaxy that resembles a loose globular cluster. Although it lies much closer than M31, it can only be seen with a large telescope because it is much smaller and fainter than M31. Although there are no more than twenty dwarf galaxies in the local group, they are mostly distributed about the Milky Way galaxy and the two large spiral galaxies, M31 and M33.

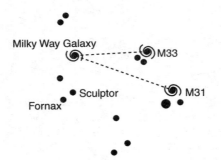

Figure 8.4(a)   The local group of galaxies

The next nearest cluster of galaxies to the local group is **the Virgo cluster** which contains over 3000 galaxies in a patch of the sky no more than about ten degrees wide. The cluster is over ten million parsecs away, of the order of twenty times further than the Andromeda galaxy. Many more clusters of galaxies have been observed at distances over 2000 million parsecs away.

ACTIVITY

Looking into clusters of galaxies

■ Triangulum M33 can be observed using a low-power telescope, on a clear night before midnight, in late Autumn in the Northern hemisphere. Although it covers as much of the sky as the Moon, its diffuse spiral arms originating from its bright nucleus are very faint. See pages 113–14 for Andromeda M31 and the Magellan Clouds.

■ The Virgo cluster includes some galaxies that may be seen with a low-power telescope in late Spring in the Northern hemisphere, although most of its galaxies are too faint to be seen except through a very large telescope. The Sombrero Hat galaxy M104 in the Virgo cluster, near the bright star Spica, is an 8th magnitude spiral galaxy seen edge on.

■ The Ursa Major cluster consists of about 12 galaxies that include two Messier objects, M81 and M82, which can be seen with a small telescope. M81 is a spiral galaxy about 36 000 light years in diameter. M82 is a smaller irregular galaxy shaped not unlike a moth – well worth looking for on a clear night!

Hubble's classification of galaxies

Edwin Hubble made a very detailed study of galaxies using the 250 cm reflector telescope in California in the 1920s. He observed the shape and structure of many elliptical and spiral galaxies and devised the classification scheme shown in Figure 8.4(b). His 'tuning fork' diagram may perhaps have been the result of an attempt to show how galaxies might evolve, but there is insufficient evidence to form any conclusions about whether or not galaxies do evolve through the different structures shown. The fact that elliptical galaxies contain a high proportion of old stars might perhaps imply the spiral arms of a galaxy are not as permanent as the central region but there is insufficient evidence for this to be anything other than an hypothesis. Indeed, there is some evidence to support the theory that an elliptical galaxy is caused by a collision between two spiral galaxies. Perhaps a more fundamental issue is why galaxies are in clusters at all. The Virgo cluster contains more than 3000 galaxies, each containing billions of stars. Instead of a single supergalaxy, the Virgo system consists of thousands of distinct galaxies. We shall return to this question in Chapter 12 when we look at what is known about the early Universe.

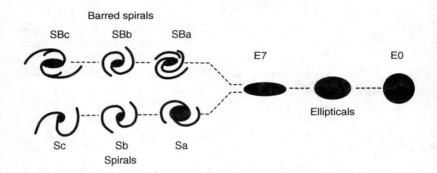

Figure 8.4(b)   Hubble's classification scheme

Elliptical galaxies

Hubble classified these galaxies on an 8-point scale according to how flat each galaxy appears to be, from E0 for a spherical galaxy to E7 which has a diameter approximately three times as long as its width. Elliptical galaxies range in size from giant ellipticals which are more than five times larger in diameter than the Milky Way to dwarf ellipticals, which are as small as a fiftieth of the Milky Way in diameter. The mass-to-light ratio M/L of an elliptical galaxy is of the order of 100 times greater than that of the Sun. We shall see later how the mass of a galaxy can be estimated. The comparatively large mass-to-light ratio of the elliptical galaxies is an indicator of the presence of hidden mass known as **dark matter** which we will return to later in this chapter.

Spiral galaxies

These galaxies were grouped into two main categories by Hubble according to whether or not the nucleus appears elongated or 'barred'. Each category is then classified on a 4-point scale, a to d, according to the tightness and patchiness of the spirals. Thus a barred spiral with loose distinct arms would be classified as an SBc spiral. The classification S0 was reserved for galaxies which appear flatter than E7 galaxies but which do not have spiral arms. The Andromeda galaxy is one of the largest known spiral galaxies. The smallest spiral galaxies are about one tenth the size of Andromeda in diameter. The mass-to-light ratio varies from twice to about twenty times the mass-to-light ratio of the Sun. As with elliptical galaxies, dark matter is present in abundance.

Dark matter

How is it possible to find out the amount of matter present in a galaxy? We will look here at two different methods which both lead to the conclusion that galaxies contain enormous amounts of dark matter. In fact, astronomers estimate that dark matter constitutes about 90% of the total mass of the Universe.

Galactic rotation

We saw on page 117 how it is possible to work out the mass of the Milky Way by using Newton's theory of gravitation to prove Kepler's third law in the form,

$$\textbf{Mass} \times \textbf{(Time period)}^2 = \textbf{(Radius of orbit)}^3$$

provided the mass is in solar masses, the time period is in years and the separation is in AU. If the time for the edge of a spiral galaxy to rotate once can be measured and the radius of rotation is known, the mass of the galaxy can be estimated.

■ The radius of rotation can be worked out from trigonometry if the distance to the galaxy is known and its angular diameter is measured. For example, if a galaxy at a distance of 12 million AU has an angular diameter of 400 seconds of arc, its diameter must be 400 × 12 million AU which is over 23 000 parsecs (since 1 parsec = 206 000 AU). The radius of rotation is then obtained by halving the diameter.

■ The time for one rotation is worked out from the circumference/speed (where the circumference = $\pi \times$ the diameter). To measure the speed, doppler-shift measurements on the light from the edge of a galaxy are made to find out the speed of rotation of stars at the edge of the galaxy.

For example, suppose the time for one rotation of the edge of a spiral galaxy is 400 million years and its radius of rotation is 4000 million AU. Using $MT^2 = R^3$, with R = 4000 million AU and T = 400 million years, gives M = 4000 million solar masses.

Exactly what form does dark matter take? This is one of the most important issues in present-day cosmology and we will return to it in Chapter 13 when we look at the future of the Universe.

Rotation rules

The speed of a star due to galactic rotation in a spiral galaxy does not depend much on its radius of rotation about the galactic centre. Measurements on the doppler shifts of stars at different distances from the galactic centre indicate little variation with distance from the centre. This is a different result from the motion of the planets round the Sun where the speed of a planet decreases the further it is from the Sun. The result indicates that there is a significant amount of hidden matter in the galactic disc compared with the galactic nucleus.

Trapped galaxies

A different method is used for elliptical galaxies in clusters, based on the fact that clusters remain as clusters even though the galaxies in them are moving about very fast. The gravitational pull of the galaxies in a cluster prevents them from dispersing. A galaxy near the edge of the cluster moving away from the cluster is pulled back into the cluster by the force of gravity of all the other galaxies, in the same way that gas molecules in the Earth's atmosphere moving upwards are pulled back by the force of gravity. Based on this simple idea, the maximum kinetic energy of a galaxy in the cluster must be equal to the work done by a galaxy to reach the edge of the cluster against the force of gravity from all the other galaxies. Using Newton's law of gravity, the maximum speed v_{MAX} of a galaxy in the cluster can be related to the total mass M of the cluster by means of the equation $v_{MAX}^2 = 2\,GM/R$, where G is the Universal Constant of Gravitation and R is the cluster radius. Using values of speed and radius, determined as outlined earlier, the total mass of the cluster can be calculated. Division by the number of observed galaxies then gives the average mass of each galaxy. The method yields an average mass which is probably more than it ought to be, on account of galaxies too faint to observe in the cluster. However, it does yield a reasonable estimate of the total mass of the cluster, which is then used to work out the mass-to-light ratio of the galaxies in the cluster.

Escape speed

The kinetic energy of an object of mass m = $\frac{1}{2}$ mv^2 where v is the speed of the object.

The work that must be done to separate completely two objects of masses m and M = GMm/R, where R is the separation of the two objects.

For a galaxy of mass m to escape from a cluster of mass M, its maximum kinetic energy must be greater than the work needed for complete separation, i.e. $\frac{1}{2}$ mv^2 > GMm/R.

The maximum speed for no escape is therefore given by $\frac{1}{2}$ mv_{MAX}^2 = GMm/R.

Rearranging this formula gives M = v_{MAX}^2 R/2G, which can then be used to calculate the total mass of the galaxy.

Measuring galactic distances

If individual stars are discernible in a galaxy, the distance to the galaxy can be measured as explained on page 110 by observing Cepheid variables or novae or, very rarely, a type I supernova. Such observations are used to determine the absolute magnitude M, so the distance modulus (m – M) and then the distance can then be worked out, if the apparent magnitude m of the star is known.

These methods cannot be used for galaxies that are too far away to see individual stars. Hubble realized that a faint galaxy is likely to be much further away than a bright galaxy. He measured the magnitude of the brightest galaxies in several clusters at known distances and worked out an average value of –21 for the absolute magnitude of these bright galaxies. Using this value, he was then able to measure the distances to clusters of galaxies, including the Virgo cluster ten million parsecs away to clusters approximately fifty times further than the Virgo cluster. An independent check on some of these measurements was possible by the simple method of comparing the apparent size of the largest galaxies in different clusters. Assuming such large galaxies are the same diameter, a galaxy ten times further than another would appear to be ten times smaller.

The distances measured by Hubble and his colleagues are truly awesome. The cluster of a dozen or so galaxies in Ursa Major lie about three million parsecs away, about four times the distance across the local group. In comparison, the much larger Virgo cluster, consisting of more than 3000 members, lies over ten million parsecs away. Yet even this enormous distance is tiny compared to the distance to the cluster of over 400 galaxies in the constellation Corona Borealis, the Northern Crown, which lies about 400 million parsecs away. Light reaching us now from this cluster was emitted over 1000 million years ago. Before we end this chapter, we will return to an even more awesome discovery made by Hubble – possibly the most important scientific discovery in cosmology of the twentieth century.

Superclusters and voids

Most galaxies belong to a cluster. By counting the total number of galaxies above a certain brightness, Hubble estimated the total number of galaxies at about 3000 million million. Even with as many as a million galaxies in every cluster, the number of clusters would be in excess of 3000 million – a number more than the entire human population of the world.

Clusters are distributed in all directions. A three-dimensional model of the distribution of all clusters at known distances reveals the presence of **superclusters** which are clusters of clusters and of **voids** which are large empty regions. In addition, clusters arranged in filaments and in sheets have been discovered. The so-called 'Great Wall' is a sheet of galaxies about 60 million parsecs away. Another very large concentration of clusters known as the 'Great Attractor' is thought to be attracting us and the Virgo cluster. However, on a larger scale, little evidence has been found for groupings and structures, and the distribution of clusters smooths out. Making measurements over distances of the order of a hundred million parsecs, the number of galaxies is the same in different directions. The distribution of galaxies can be likened to the distribution of material in a sponge. The holes represent voids and the sponge material represents galaxies. The distribution of sponge material is very uneven on a 'hole size' scale but it is uniform on a much larger scale. In 1999, an Anglo-German team of astronomers confirmed this picture after mapping the Universe in infra-red radiation, up to 100 megaparsecs away.

- Galaxies are between 1 and 200 kiloparsecs in diameter at separations of the order of 1 million parsecs.
- Clusters and superclusters range from about 200 kiloparsecs to about 10 million parsecs in diameter.
- Over distances greater than about 100 million parsecs, clusters are distributed evenly.

ACTIVITY

A cluster distribution model

Scatter a handful or two of grass seeds thinly over a centimetre grid at random. You ought to observe that the seeds are not evenly distributed. Some squares will be more densely populated with clusters of seeds than other squares. However, the number of seeds on equal rows of squares of sufficient length in different parts of the grid should even out. A measure of the evenness of the distribution could be worked out from the ratio of the average width of a cluster to the length of a typical row of 'even' distribution. A measure of 1 would mean that clusters are distributed evenly. A measure of 0.1 corresponds to uniformity on a scale ten times larger than a typical cluster. In these terms, the distribution of clusters of galaxies on a one-dimensional scale is of the order of 0.02, corresponding to uniformity on a scale about 50 times the size of a typical cluster. In terms of three dimensions, this unevenness would be about 10^{-5} (= 0.000001 = 0.02^3 approximately).

Hubble's great discovery

Red shift

The speed of a distant galaxy can be measured from the doppler shift of its spectrum. As explained on page 82, the lines of the spectrum of light from a star are shifted due to the star's motion if it is moving towards or away from us. The spectral lines are shifted to longer wavelengths (i.e. a red shift) if the light source is receding, and to shorter wavelengths if the light source is approaching (i.e. a blue shift). By observing the spectrum of light from different galaxies using a 60 cm telescope at the Lowell Observatory, Vesto Slipher in 1917 discovered that certain spiral galaxies are moving away from us at speeds of more than 500 km/s, much faster than any object in our own galaxy. The term 'red shift' was coined for the ratio of the change of wavelength to the emitted wavelength.

Hubble's law

Edwin Hubble followed up Slipher's work by estimating the distances to two dozen galaxies of known redshifts within two million parsecs of the Milky Way galaxy. His results, published in 1929, showed that the red shift increased with distance. Further measurements of more galaxies were made by Milton Humason, using the 250 cm telescope at Mount Wilson. By 1935, Hubble and Humason were able to publish the results for more than 140 galaxies, out to distances of more than 300 million parsecs, moving away at speeds over 40 000 km/s. The results confirmed Hubble's findings of 1929 that the red shift increased with distance. More importantly, by plotting the results on a graph of red shift against distance, it was clear that the red shift and hence the speed of recession is in proportion to the distance. This relationship is known as **Hubble's law**. The constant of proportionality in the relationship is known as the Hubble constant, H.

Speed of recession = the Hubble constant × distance
υ H   d

Hubble and Humason estimated the Hubble constant to be 530 km/s per million parsecs. The unwieldy unit is usually written as $km\ s^{-1}\ Mpc^{-1}$. Subsequent measurements using bigger telescopes and improved detectors have reduced the Hubble constant to a present-day value of about $65\ km\ s^{-1}\ Mpc^{-1}$.

The value of H has a very profound significance as it is used to estimate the age of the Universe. We will consider this in detail later in this chapter and again in Chapter 10. Hubble's law is an experimental law valid for a limited range of measurements. Its possible explanations caused great controversy for half a century after it was announced. It is now accepted that Hubble's law follows because the Universe is expanding from a primordial explosion billions of years ago. This explosion, known as the

Big Bang, was the origin of space and time. Not surprisingly, this theory had many opponents in the scientific community. Its advocates developed Einstein's ideas on gravity, space and time to show that an expanding Universe is mathematically consistent with Einstein's theories. Nevertheless, in spite of Einstein's stature, the Big Bang theory was vigorously opposed until evidence from several independent experiments was found. The Universe is expanding and the distant galaxies are moving away from each other. The further away a galaxy is, the faster it moves away. We will look in more detail at Einstein's ideas on space and time before we return to the Big Bang in Chapter 10.

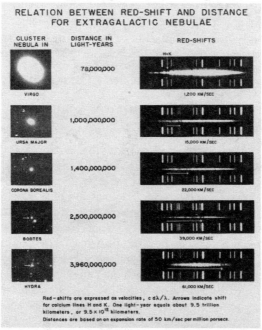

Figure 8.5(a)   Sample evidence for Hubble's law
(© Carnegie Institution of Washington)

The spectrum of light for each galaxy is the narrow horizontal strip of light between each pair of identical laboratory spectra. Two prominent lines, the calcium H and K lines, are evident in each strip. The red shift is indicated by a horizontal arrow below each strip. The photograph of each galaxy shows that the smaller the galaxy appears to be, the further away it is.

Figure 8.5(b) Edwin Hubble (1889–1953) (Science Photo Library)

Hubble abandoned a promising career in law to become an astronomer. He was an outstanding experimenter with a high degree of intuition. He commissioned the Mount Palomar 500 cm telescope in 1949 and was the first person to use it.

Quasars

In 1954, Cygnus A was the first galaxy to be identified as a strong source of radio waves. Within a few years, many more radio galaxies had been detected. By measuring the red shift of a galaxy, its speed and therefore its distance can be determined. By 1960, radio galaxies with red shifts almost as large as 0.5 corresponding to a 50% increase of wavelength had been detected. As radio telescopes became more sensitive, radio astronomers were able to detect more and more radio sources beyond the Milky Way galaxy. In 1962, Cyril Hazard in Australia managed to identify a particular radio source, 3C 273, as a 13th magnitude star. Later that year, Maarten Schmidt used the large telescope at Mount Palomar to obtain the spectrum of this source and discovered an unusually large red shift of 0.15, corresponding to a speed of about 0.15 c, where c is the speed of light in space, and a distance of about 900 Mpc. Its absolute magnitude is therefore about –27 which makes it about 1000 times as luminous as the Milky Way galaxy. This discovery made 3C 273 even more powerful than many galaxies yet it appeared more like a star as a point rather than an extended object like a galaxy.

Astronomers worldwide were astounded by Schmidt's discovery. More similar star-like objects with very unusually large red shifts were discovered, usually on the basis of excessive ultraviolet radiation. The unwieldy term 'quasi-stellar object', subsequently shortened to **quasars**, was coined for these sources. By 1965, quasars with red shifts as large as 2 had been discovered. Twenty years later, the red shift limit had been pushed over 4, corresponding to 0.88 c. In addition, detailed radio maps indicated the presence of fast-moving clouds and jets of matter. Intensity variations on time scales no larger than years, and as small as a few minutes, indicated that these objects were no larger than a few light years, much much smaller than even the smallest galaxies. What could be the cause of energy being radiated on a galactic scale from objects not much larger than our Solar System? Neither nuclear fusion nor nuclear fission can release energy per unit mass of fuel on such an enormous scale.

Galaxies with distinct bright point-like centres or 'active galactic nuclei' were first observed by Carl Seyfert two decades before the discovery of quasars. Astronomers quickly recognized that quasars might be distant galaxies with active centres. The form of electromagnetic radiation from quasars was identified as similar to the radiation produced by high-speed charged particles in a synchrotron accelerator which is essentially a large evacuated tube in a ring of magnets. The charged particles are accelerated by electrodes at intervals along the ring as they are forced round the tube by the magnets. Images of jets of material emitted from quasars reinforced the theory that matter spiralling round the centre of an active galactic nucleus is dragged in, releasing huge amounts of energy in the form of a jet of radiation as it falls into the vortex. Estimates of the mass of galactic nuclei made using the methods outlined on page 117 indicate masses of the order of thousands of millions of solar masses. The idea that there might be a massive **black hole** lurking at the centre of such a galaxy offered one possible way of explaining how so much energy can be produced in such a small region. Black holes are a theoretical outcome of Einstein's ideas on space and time. Not even light can escape from a black hole. An object falling into a black hole radiates energy before it disappears for ever into the black hole. The theory that a quasar is caused by a black hole is attractive, since it accounts for the small width of the quasar, and for its enormous rate of energy production. Perhaps there is a massive black hole at the centre of the Milky Way galaxy, directing jets of radiation far into space. We will return to the ideas and evidence for black holes in the next chapter.

Red shift reminders

Working out the speed from the red shift

The theory of the Doppler effect for a receding source moving at speed υ relates the observed wavelength λ to the emitted wavelength λ_e in accordance with the equation $\lambda = \lambda_e (1 + \upsilon/c)$ provided υ is much less than the speed of light c.

Relativistic time dilation must be taken into account when υ is not much less than c. In addition, the rate of expansion of the Universe must be considered, as we shall see in Chapter 12. By definition, the red shift z = change of wavelength/emitted wavelength $= \dfrac{\lambda - \lambda_e}{\lambda_e}$

The table below gives the red shift z for different values of υ/c, taking account of time dilation.

υ/c	0.00	0.10	0.20	0.30	0.40	0.50	0.60	0.70	0.80	0.90
z	0.00	0.10	0.22	0.36	0.53	0.73	1.00	2.40	3.00	4.36

Working out the distance d from the speed

Hubble's law υ = Hd enables the distance d to be worked out from the speed, given a value for H and the information that the speed of light c = 300 000 km/s. For example, for H = 50 km s^{-1} Mpc^{-1}, if υ = 0.1 c = 30 000 km/s, then d = υ/H = 30 000/50 = 600 Mpc.

Working out the absolute magnitude from the observed magnitude m and distance d

This is explained on page 70. See Figure 6.2 for the conversion table.

Summary

Distance measurements

- **nearby stars**: parallax method
- **distant stars**: use distance modulus and main sequence fitting or Cepheid variable method or nova estimates
- **nearby galaxies**: Cepheid variables, nova and supernova estimates
- **galaxies too far away to resolve individual stars**: apply distance modulus to brightest in a cluster or to a supernova or estimate distance from width or measure red shift and assume Hubble's law

Types of stars and galaxies

- **Cepheid variables**: variable stars with variability periods from days to weeks
- **Population I stars** are hot, blue, metal-rich and young, to be found in the spiral arms of galaxies
  Population II stars are red giants, metal-deficient and old, to be found in globular clusters and in the galactic centre
- **Hydrogen atoms** in space emit radio waves at 21 cm, mostly from dust and gas in spiral arms
- **The Sun** lies in the Orion arm of the Milky Way galaxy. The galactic centre lies beyond the Sagittarius arm of the Milky Way galaxy
- **Hubble's classification of galaxies**: S(B) a–d for spiral galaxies with B if the centre is bar-shaped; E0–E7 for elliptical galaxies
- **Galaxies are in clusters** which are often in clusters of clusters (superclusters) with voids between; 90% of the mass of a galaxy may be dark matter
- **Clusters are distributed evenly** over distances greater than about 100 million parsecs
- **The red shift** of a receding galaxy is a measure of the increase in the wavelength of light emitted by the galaxy
- **Hubble's Law** states that the speed of recession of a galaxy = the Hubble constant × distance
- **A quasar** is thought to be a distant active galactic nucleus where a massive black hole is attracting surrounding matter

9 | BLACK HOLES

In the previous chapters, we have met some very unusual astronomical objects such as supernovae, pulsars, quasars and black holes. The observational evidence that led astronomers to these objects has been presented in some detail to reinforce the general philosophy that underlines modern science, in particular the link between observation and theory. No amount of evidence from observations or experiments can ever prove a theory but just one valid piece of scientific evidence is enough to disprove it. In this chapter, we will look at some of the ideas in current theories about gravity, space and time that explain such objects as pulsars, quasars and black holes. We will look at some of the predictions that follow from current theories and the experimental evidence available that supports these predictions. For example, the prediction that light is deflected by gravity follows from Einstein's ideas about gravity. Evidence to support this prediction has been found in experiments involving radar and space probes and in astronomical observations of double images of distant galaxies. Such support boosts the credibility of other predictions derived from the work on gravity by Einstein and others, notably the Big Bang theory of the origin of the Universe.

In this chapter, we will concentrate on ideas, predictions and evidence, touching only lightly on the mathematical tools used to turn ideas into predictions in this field of study. If the ideas seem to be almost beyond credibility, it is worth remembering that we are following in the footsteps of scientists such as Einstein. The top of a mountain is just as high, regardless of how many times it has been reached before.

Gravity and light

After Newton put forward his theory of gravity, a tale was told of how his thoughts on gravity originated when an apple fell on his head as he was sitting under a tree in summer at his Lincolnshire home. Regardless of the veracity of the tale, it conjures up a memorable picture of an object being attracted by the force of gravity due to the Earth. It neglects the question of how the force of gravity acts between two objects. Essentially, Newton's theory of gravity established that there is an attractive force between any two objects, and that the size of the force depends on the masses of the objects and on their distance apart, as explained on page 40. Newton introduced a constant of proportionality, the universal constant of gravitation G, which he assumed had the same value at all points in space and at all times.

Newton's law of gravity with its 'inverse square relationship' linking the force of gravity between two objects to their separation, successfully explains many astronomical observations and events, including the motion of the planets and comets round the Sun, of moons round the planets, of satellites round the Earth, and of the tides on the Earth. It is used to predict tides, eclipses, comet and satellite orbits, planetary positions, escape speeds from planets, planetary atmospheres and many other astronomical and planetary features. Yet all the predictions that follow from Newton's law assume a physical property which was challenged and tested by Galileo, and which became the focus of fundamental ideas about gravity developed by Einstein early in the twentieth century.

From Galileo to Einstein

Science can be very entertaining, as Galileo reportedly showed in Pisa when he demonstrated that falling objects released simultaneously from the top of the Leaning Tower of Pisa hit the ground at the same time. He had already shown that the acceleration of a falling object is independent of its weight. Even today, this finding comes as a surprise. Many people generally imagine that a heavy object ought to descend at a faster rate than a lighter object because the force of gravity on the heavier object is greater. Newton provided the explanation when he showed that the acceleration of any object acted on by a force is:

1 proportional to the force on the object, and
2 inversely proportional to the mass of the object.

For example, the acceleration of a lorry from standstill will be reduced if the lorry is carrying a heavy load even though the engine force is the same. In the case of a falling object, the acceleration is unaffected by the mass. For example, if the mass is doubled, the force of gravity is doubled too but this increased force acts on double the mass so the acceleration is unchanged. The increase in the force of gravity on a larger mass in free fall has no effect on the acceleration because the force is acting on an increased mass.

ACTIVITY

Drop two different-sized coins at the same time from the same height above a level floor. The two coins should hit the floor simultaneously and so the sound of their impacts is heard as one sound only.

The mass of an object determines the force of gravity on the object. Two objects that experience equal forces of gravity must have equal masses. If the force of gravity on an object is ten times the force of gravity on a different object, the masses of the two objects must be in the ratio ten to one. The mass of an object can be compared with the mass of a different object using a weighing machine. If the two objects have equal weights, they must have equal masses.

Another property of mass is that it gives an object inertia. Inertia is a measure of the difficulty of causing change in the velocity of an object. A supertanker at sea has a huge inertia and takes several kilometres to stop just because it has a very large mass. The same would apply to a rocket in space, well beyond the force of gravity of the Earth, or any other body. Changing the velocity of an object is more difficult the larger the mass of the object. So one method of comparing the masses of two objects is to compare how they respond to the same force. If two objects have the same mass, they should take the same duration to reach the same speed when acted on by the same force.

The masses of two objects can be compared by weighing them, or by comparing their inertial response to the same force. If the mass of one of the objects is known, the mass of the other object can be calculated by either method. The gravitational mass (i.e. mass measured by the weighing method) should be equal to the inertial mass (i.e. the mass measured by changing its velocity). This has been tested by different experimenters, using different methods, employing sophisticated equipment to give the same result to the astonishing degree of accuracy of less than 1 part in 100 million. The overall conclusion is that the ratio of the gravitational mass of an object to its inertial mass is the same for any object. The equality of the gravitational mass to the inertial mass of an object is the reason why falling objects descend at the same rate.

Since gravitational mass is equal to inertial mass, is it possible to distinguish between gravity and accelerated or decelerated motion? In Chapter 5, we saw that it is impossible to distinguish between being in a state of rest and moving at constant speed without change of direction. What about distinguishing between gravity and changing motion? Here are some situations to help you decide.

■ Imagine you are in a descending elevator (lift in Britain) moving at steady speed. You cannot tell that it is moving if you can't see outside the elevator and it is moving exceedingly smoothly. However, if the elevator suddenly stops, you feel extra heavy, as if your weight was greater than normal. This is because the extra force of the elevator floor on you stops you descending. In a completely enclosed elevator, is it possible to tell if your extra heaviness is because the lift suddenly stopped, or because the earth's gravity has momentarily increased?

■ Now imagine you are floating about in a space station in orbit above the Earth. Moving about is too easy if you are not fastened in a seat. The problem is stopping yourself bouncing about in the space station. What has happened to the gravity that kept your feet on the ground on the Earth? The fact is that the force of gravity on you is still present, but it is being used to keep you whirling round the Earth. You feel weightless as if gravity has been cancelled out inside the space station. If the space station were a gigantic donut-shaped tube, artificial gravity could be created as a result of making the space station rotate at a suitable speed. You would be held against the wall of the tube, which would then be like a floor.

■ Now back to the elevator. Suppose the elevator cable snaps and you descend in free fall. You can float about without the need for support when in free fall, as if you were in a gravity-free environment. In effect, the Earth's gravity is cancelled out by the downward acceleration of you and the elevator.

Einstein thought very deeply about the link between gravity and accelerated motion. A person floating about in a windowless capsule would not know if the capsule was in a gravity-free environment far away from the Earth or in free fall above the Earth – until the capsule hit the ground! He came to the conclusion that the effects of gravity and accelerated motion are identical and are in effect indistinguishable. On this basis, known as **the Principle of Equivalence**, he developed his General Theory of Relativity in which he proved that a gravitational field causes space and time to be distorted.

How to bend light

Light travels on a straight path – unless it passes through a strong gravitational field. We will look at the actual experimental evidence that gravity bends light later in this section. Einstein worked out how much a light beam passing near the Sun should be bent by the Sun's gravity. His calculations were based on very complicated mathematics. The essential idea is not too complicated if we start by thinking about a light beam passing through opposite-facing portholes of an accelerating rocket. Figure 9.1(a) shows the idea.

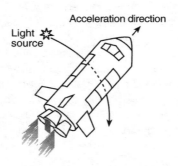

Figure 9.1(a)   Light bending in an accelerating rocket

If the path of the beam could be made visible to an observer in the rocket, perhaps by making the beam skim across a screen, the observer would see a curved path like the path of a stone thrown sideways from the top of a tower. The acceleration of the rocket causes the observer to see a curved light path. As acceleration cannot be distinguished from gravity, it therefore follows that gravity should cause a light beam to curve too.

Einstein worked out that a light beam that passes near the edge of the Sun should be deflected by an angle of 1.75 seconds of arc. He proved that the angle of deflection is equal to $\frac{4GM}{Rc^2}$, where M is the mass of the Sun, R is its radius, G is the universal constant of gravitation and c is the speed of light. Einstein's prediction from his General Theory of Relativity was tested in 1918 by a team of astronomers, led by Arthur Eddington, who observed and measured the deflection of stars occulted (hidden) by the Sun during a total solar eclipse. Einstein's predictions agreed with Eddington's measurements closely enough to give a huge boost to Einstein's General Theory of Relativity.

Gravitational lenses

If gravity bends light, is it possible to make gravity form images? A magnifying glass is a glass lens that makes objects appear larger. It does this by changing the direction of light passing through it, making the light appear to have come from an enlarged image.

A straight-sided glass of water can act as a magnifying lens too. Look at a pencil through a glass of water and you will see a magnified image of the pencil, probably distorted if the pencil is very close to the glass. You might even see an extra image at each edge of the glass, as well as the central image. The extra images are formed because light from the pencil travelling to the edges of the glass is bent by the combined effect of the water and the glass, as shown in Figure 9.1(b). Can gravity cause distorted images of objects in space?

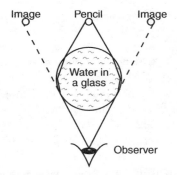

Figure 9.1(b)   Light bending to form a double image

In 1979, a double quasar Q0957 and 561 was discovered. Because the signal from each part fluctuated in the same way, it was realized that the two parts were in fact two images of a single quasar, caused by an intervening galaxy or a cluster. The quasar is hiding behind this very large mass, but we see the quasar because its light skimming the edge of the mass is bent round, just as light from the pencil in Figure 9.1(b) is bent round each edge of the glass.

More recently, the Hubble Space Telescope has revealed images of faint galaxies distorted and spread out into streaks behind clusters which act as enormous lenses. By combining measurements and theory, the mass of an intervening cluster or galaxy can be calculated and the distance to the quasar can be worked out. See Plate 6.

ACTIVITY

The pencil test

Observe an object such as a pencil through a glass of water, as in Figure 9.1(b). Place a card in front of the glass to cover the middle part. You ought to be able to see an image of the pencil either side of the card. Remove the card and observe your finger through the glass, touching the other side of the glass. The distortions you can see vary according to where the point of contact is.

Gravitational red shift

If light is bent by gravity, does its energy change when it enters or leaves a gravitational field? An object thrown into the air loses speed as it rises and therefore loses kinetic energy. What about light directed upwards? As explained on page 51, light consists of photons, which are wavepackets of electromagnetic energy. The energy of a photon is in proportion to its frequency. Does the energy of a photon change in a light beam directed upwards? According to Einstein's theory, the energy of a photon ought to decrease if it moves up because it uses some of its energy to overcome gravity. This decrease of energy causes a decrease of frequency, which is the same as an increase of wavelength or a 'red shift'. Einstein worked out that the gravitational red shift for each metre of height gained near the Earth's surface is 10 Hz for an ultraviolet photon of frequency 1×10^{17} Hz. This prediction was tested at Harvard University by Pound and Rebkha in 1959, who measured the frequency shift of gamma photons as a result of travelling up a 22.5 m tower from the radioactive source to a detector at the top of the tower. Their result agreed with Einstein's prediction to within one per cent. In a further test in 1976, an atomic clock flown to high altitude and back in a rocket was compared with an identical atomic clock on the ground. An atomic clock contains atoms that vibrate at a precise frequency. The operation of the clock is based on the vibrations of these atoms just as the operation of a pendulum clock is based on the vibration of a pendulum. The clock in the rocket ran slower than the other clock when above the Earth's surface, so its reading fell behind the other clock. The results agreed with Einstein's predicted frequency shift to within 1 part in 100 000.

Calculating the gravitational red shift

Einstein proved that the ratio of the change of frequency to the initial frequency is equal to gh/c^2, where c is the speed of light, h is the height gain and g is the strength of gravity. Near the surface of the Earth, g = 9.8 newtons per kilogram so $gh/c^2 = 10^{-16}$ for each metre rise.

About black holes

In the previous section, we saw that light is deflected by a strong gravitational field. In addition, the frequency of a light photon or an atomic clock is affected by gravity. A gravitational field distorts space and time. In his General Theory of Relativity, Einstein showed that a gravitational field acts by curving space and time. One of the most dramatic effects predicted by Einstein's theory is that light cannot escape from an object which creates sufficiently strong gravity. Such an object is referred to as a **black hole**.

These fascinating objects were first predicted by Reverend John Michell, in 1783, although he did not invent the name itself. In fact, the term 'black hole' is of much more recent origin, dating from 1968 when it was coined by an American physicist, John Wheeler. Michell knew from Newton's theory of gravitation that an object projected directly upwards would escape from the Earth if its speed exceeded 11 km/s. This speed is referred to as the **escape speed** from the Earth. The escape speed from the surface of a planet or a star depends on the strength of gravity at the surface. The stronger the gravitational field, the greater the escape speed needs to be. Michell reckoned that if the mass of a planet or star were great enough, not even light travelling at a speed of 300 000 km/s could escape from its surface. Michell even worked out that a star would need to be about 500 times as large as the Sun to prevent light escaping.

Shortly after Einstein published his General Theory of Relativity, Karl Schwarzschild derived an exact solution to Einstein's equation linking space, time and gravity. He proved that the radius of a black hole, known as the **Schwarzschild radius R**, is related to its mass M in accordance with the equation $R = \frac{2GM}{c^2}$. If the Earth contracted in diameter to become smaller and smaller without loss of mass, it would need to shrink to about 18 millimetres in diameter before it became a black hole. In comparison, a contracting galactic nucleus of a thousand million stars would become a black hole when all the stars fitted into a space no larger than the orbit of Pluto.

Journey into the unknown

The Schwarzschild radius defines the **event horizon** of a black hole. This means that any object closer than the Schwarzschild radius to the centre of the black hole is trapped for ever inside the event horizon. It would be impossible for an outsider to see an object inside the event horizon as no light from the object could pass through the event horizon. The crew of a spaceship being pulled towards a black hole might not be aware that they had passed through the event horizon as they could still see light from distant stars.

What happens to the crew of the spaceship once they are inside the black hole? According to Schwarzschild, the spaceship is drawn inexorably towards the centre, to be pulled to shreds by a gravitational field which becomes ever stronger the closer the spacecraft approaches the centre. The spaceship and the crew, or their remains, would be stretched more and more as they are pulled into the centre, sucked in like strands of spaghetti. Perhaps this awful fate awaits any inhabitants of planets at the centre of the galaxy where astronomers reckon a black hole lurks, feeding on surrounding stars to become heavier and ever more powerful. Could the Sun and the planets end up in such a cosmic monster?

If nothing can escape from a black hole, it would seem that a black hole is destined to swallow up matter for ever. This prospect conjures up disastrous possibilities, but worse is to come in the form of the **spinning black hole**, dragging space and time round near its event horizon. This was 'discovered' in 1963 by Roy Kerr as an outcome of Einstein's equation. When a massive body collapses, any object near the event horizon would be subjected to a dragging effect due to the rotation of the black hole, which could pull it through the event horizon into eternal oblivion. However, an important property of a rotating black hole is that energy can be extracted from it. For this to happen, an object near the event horizon is broken in two so one part is flung into the black hole against its rotation while the other half is catapulted into space by the rotating motion of the black hole. The first part reduces the black hole's rate of rotation, reducing the energy of the black hole, while the second part carries away kinetic energy into space. Perhaps this is the mechanism which enables a black hole at the centre of a galaxy to generate its awesome power.

Black hole rules

A black hole formed by a massive collapsing star is thought to have no identity other than its mass (which determines its Schwarzschild radius) and its rotation (technically referred to as its angular momentum). A distant external observer cannot tell what object or objects the black hole formed from. This is known by the curious name of the **'no hair' theorem**, as any attempt to attach anything to a black hole is doomed to failure. Since the surface area of the event horizon of a black hole increases with increased mass, this area is a measure of how much information has been lost through the event horizon. Loss of information creates disorder, which in scientific language is referred to as **entropy**. Disorder, or entropy, always increases in any isolated physical process. This is why we can't recover all the energy we obtain from fuel and use it again. In the same way, black hole processes in an isolated system create disorder and increase the entropy of the system.

By using analogies with thermodynamics, certain rules for black holes have been developed which make them less likely to create catastrophe. Stephen Hawking at Cambridge has shown that a black hole can actually evaporate as a result of swallowing up antiparticles from pairs of particles and antiparticles created near the event horizon. For every antiparticle swallowed up, its particle counterpart is emitted to form part of a stream of radiation away from the black hole. The mass of the black hole is steadily reduced as a result of swallowing antiparticles at the same time as it emits a stream of radiation. The energy of this radiation is a measure of the **temperature** of the black hole. The smaller the mass of the black hole, the higher its temperature and the quicker it evaporates. Hawking worked out that the time taken to evaporate is longer than the age of the oldest stars if the mass is more than about 10^{12} kg. The Schwarzschild radius corresponding to this mass is about the same diameter as the nucleus of an atom. A black hole at the centre of a galaxy is much more massive, and will be there for a long long time, dragging in surrounding matter which emits high-energy radiation as it accelerates towards the event horizon.

Evidence for black holes

The two examples below could be explained by means other than black holes. However, black holes offer the most convincing explanations to date of the observations.

- **Cygnus X1** is an X-ray emitting object of about 6 solar masses in a binary system. The object could be a black hole drawing matter from its companion. According to this theory, X-rays are emitted from this matter as it is accelerated towards the event horizon. Other similar X-ray sources have been detected. The X-rays from these sources and from Cygnus X1 fluctuate rapidly, indicating such sources are no larger than a few kilometres in diameter, perhaps caused by the collapse of a star more than twenty times as massive as the Sun.
- **M87** is a giant galaxy with a very bright active nucleus (see Plate 7). Part of the galactic centre is spinning so fast that it is thought to contain about 3000 million stars in a region no more than the size of the Solar System. In Chapter 12, we will return to how such an object might have formed.

Ideas beyond imagination

After publishing his Special Theory of Relativity in 1905, Einstein worked on his ideas linking gravity, space and time until he published his General Theory of Relativity in 1916. Much of his work involved very complicated mathematical methods, which he combined with insight and intuition that took him well beyond the level at which any of his contemporaries were working. Just as Newton over two centuries earlier had taken mathematics and physics forward in great leaps, Einstein was to do likewise to produce a theory whose predictions were completely unexpected and totally beyond the visions of any other scientist or mathematician since Newton. We will look later in this chapter in detail at the experimental evidence that supports the General Theory. For the moment, however, let's look briefly at how the theory emerged and the ideas that underpin it.

In the Special Theory, Einstein had shown that space and time are not absolute quantities, and he derived the equations in Chapter 5 that relate observations made by a stationary observer to those made by an observer moving at constant velocity. In effect, his equations enable the laws of physics, such as conservation of energy, to be identical in all frames of reference moving at constant velocity relative to each other. Between 1905 and 1916, he worked on trying to show that the laws of physics are the same in accelerated frames of reference. He recognized the significance of the link between accelerated motion and gravity and went on to show how gravity affects space and time.

To appreciate some of Einstein's work, a good place to start is a map of your home country on a flat table. Suppose you want to work out the direct distance between your home and another location on the map which is 80 km north and 50 km east of your home. The direct distance can be worked out using Pythagoras' theorem.

The direct distance s is given by the equation $s^2 = x^2 + y^2$ where $x = 80$ km and $y = 50$ km.

$$\therefore s^2 = 80^2 + 50^2 = 8900 \qquad \text{so } s = \sqrt{8900} = 94 \text{ km.}$$

Now suppose you want to find the direct distance along the surface between two points far apart on a globe of the Earth. Pythagoras' theorem won't work on this scale, because the distance is not insignificant compared with the globe radius. Einstein knew from the work of mathematicians, such as Gauss in the nineteenth century, that the distance δs between any two nearby points on a curved surface can be expressed in a general form,

$\delta s^2 = g_{11} \delta x_1{}^2 + g_{12} \delta x_1 \delta x_2 + g_{22} \delta x_2{}^2$ where δ signifies a small distance and x_1 and x_2 are the two coordinates of any point on the surface, for example the distance along the equator and the distance north or south of the equator. Each of the quantities g_{11}, g_{12} and g_{22} changes its value with position, because the surface is curved.

The metric tensor

A point in space and time is defined by three spatial coordinates (x_1, x_2 and x_3) and a time coordinate which we will label x_4. The formula above can be extended to any two nearby points in space and time to give the following equation, which is known as the **metric**, in which g_{ij} is referred to as the *metric tensor*.

$\delta s^2 = g_{ij} \delta x_i \delta x_j$ where i and j run through all the possible combinations of 1, 2, 3 and 4.

The quantity δs represents the separation between two points in space and time. This quantity is the same regardless of the system of coordinates chosen. The metric tensor, g_{ij}, however varies with position, like the strength of the Earth's magnetic field does. A tensor is any quantity that transforms from any one coordinate system into any other coordinate system according to some basic rules. By expressing the laws of physics in the form of tensors, Einstein was able to show that the laws are independent of the coordinate system used to describe them. He worked out certain other tensors which are derived from the metric tensor and discovered that one particular tensor, now known as the Einstein tensor E_{ij}, determines the curvature of space. As outlined in Appendix 4, he was able to relate E_{ij} to the distribution of matter in space and deduced the following law of gravitation that describes the empty space around a massive object:

$$E_{ij} = 0$$

This equation is astonishingly easy to express yet exceedingly difficult to solve. Within two months of the publication of the General Theory of Relativity, Karl Schwarzschild had solved it and derived formulas for each component of the metric tensor. In particular, he showed that the component for the distance coordinate from the centre of the massive body of mass M included the factor $\frac{1}{(1 - 2GM/rc^2)}$. The implication of this factor was immediately obvious to Schwarzschild, Einstein, Eddington and many other scientists. Something very very odd happens at $r = 2GM/c^2$, now known as the Schwarzschild radius, because this value of r makes the bottom line of the above factor zero which makes the factor itself infinitely large. Working out the effect of this factor on an outgoing light ray, it became clear that such a light ray is trapped within a sphere of radius $2GM/c^2$, and this sphere is an event horizon that prevents anyone outside seeing inside. The idea of a black hole became mathematically respectable. Schwarzschild also predicted the existence of a singularity at the centre, a point at which anything inside the event horizon ends its existence.

Other predictions of Einstein's General Theory of Relativity included:

- the gravitational red shift of a light ray emitted from a point outside the event horizon,
- the deflection of a light ray by any gravitational field,
- the precession of the orbit of a planet.

We will look in detail at the observational evidence that supports these predictions in the next section. The discoveries in astronomy of supernovae, quasars, neutron stars, pulsars, black holes and other astonishing objects rekindled interest in general relativity. As mentioned earlier, in 1963, Roy Kerr discovered another solution to Einstein's law of gravitation – the rotating black hole. He showed that Schwarzschild's solution is a special case of zero rotation. Current research by Stephen Hawking and others is attempting to show how gravity relates to the other forces of nature and to find out the effect of gravity on the evolution of the Universe. We shall return to this fascinating question in Chapter 10.

Towards general relativity

Einstein's search for a new theory of gravity began in 1907 with his work on linking special relativity to accelerated motion and to gravity. He worked out the effect of gravity on light and by 1912 he realized the need to use tensors to describe the curvature of space-time. A mathematician colleague, Marcel Grossmann, a friend from his student days, encouraged Einstein to study the properties of tensors but Einstein did not think he could generate Newton's law of gravitation by this method. He attempted unsuccessfully to find alternative approaches for three years until 1915 when he returned to his friend's ideas. After more than a year working on this approach, he was able to publish the General Theory of Relativity.

Testing times

When Einstein published the General Theory of Relativity in 1916, he also showed how it explained a problem about the orbit of the planet Mercury that had puzzled astronomers since 1859. Mercury's orbit round the Sun is elliptical, and its distance from the Sun varies from 57.4 million kilometres at perihelion (i.e. at its nearest to the Sun) to 58.6 million kilometres at aphelion, its furthermost point from the Sun. Since it takes 88 days to complete each orbit, it reaches perihelion once every 88 days. Careful measurements made on its orbit revealed that its perihelion was advancing gradually at a rate of 0.159 degrees (= 574 seconds of arc) per century. Le Verrier worked out that most of this advance could be explained by the effect of the other planets on Mercury. However, he was unable to account for 43 seconds of arc and, for half a century, nor could anyone else – until Einstein! Using his theory to modify the equation from Newton's laws for the orbit, Einstein showed that an advance of the perihelion in the absence of any other planet was to be expected. He derived a formula for the expected advance per orbit and used it to calculate that Mercury's perihelion ought to advance at a rate of 43 seconds of arc per century – exactly the amount that Le Verrier in 1859 had been unable to account for. This result astonished scientists worldwide and gave an immense boost to the credibility of Einstein's General Theory.

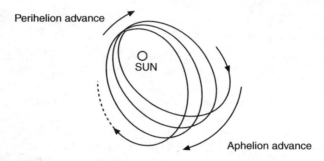

Figure 9.2   Mercury's orbit

Another of Einstein's predictions from the General Theory of Relativity is that light grazing the Sun is deflected by 1.75 seconds of arc. He predicted half this deflection in 1911 before he began to develop his ideas using tensors. When he published the General Theory in 1916, he showed that his equation predicted an angle of deflection equal to $\frac{4\,GM}{Rc^2}$, where M is the mass of the Sun, R is its radius, G is the universal constant of gravitation and c is the speed of light. This gave a value for the Sun of 1.75 seconds, twice his earlier prediction. The outstanding success of his explanation of the perihelion advance of Mercury's orbit led the foremost astronomer of the age, Sir Arthur Eddington, to plan an expedition to South America to observe the total solar eclipse of 1919, in order to measure the shift of position of stars as they were occulted by the solar disc during a total solar eclipse. The result confirmed Einstein's prediction to an accuracy of about 30%.

The success of the gravitational redshift tests described on page 142 in producing observations in agreement with Einstein's predictions is not now reckoned as convincing evidence for general relativity. This is because gravitational red shifts can be explained in terms of the equivalence between gravity and accelerated motion, without the necessity of using general relativity. The fact that general relativity does predict the red shift does not mean that general relativity is the only explanation.

Convincing evidence for general relativity rested only on the deflection of light near the Sun until recent decades. Between 1968 and 1978, measurements were made of the transit time of reflected radar pulses sent from Earth to Mercury, Venus and Mars and to the Viking and Mariner space probes. The measurements were made as each object was eclipsed by the Sun. The measurements proved to within 0.1% that the deflection was 1.75 seconds of arc. There is now little doubt that space is curved near the Sun, and the curvature is as predicted by Einstein.

Gravitational radiation

Einstein predicted that binary stars should emit gravitational waves as they orbit about each other. If gravity causes space to be curved, the stars in a binary system should create vibrations in the curvature of space. General relativity predicts that gravitational waves travel through space at the speed of light. Convincing evidence for such waves was discovered by Russell Hulse and Joseph Taylor in 1974 as a result of searching for radio pulsars. As explained on page 103, a pulsar is a fast-spinning neutron star, formed as a result of a supernova when a massive star explodes. Hulse and Taylor were using the giant radio telescope at Arecibo in Puerto Rico to search for radio pulsars when they discovered a binary pulsar PSR 1913 + 16, which proved to be very unusual. The pulsar itself emits its radio pulses at a constant rate of 59 milliseconds as it and its companion move round each other. Several other binary pulsars had been detected but not one which had an orbit no larger than the Sun and took just 8 hours for each orbit. The speed of the pulsar on its orbit worked out to be 300 kilometres per second. In such a close orbit, the perihelion (i.e. position of least distance apart) advances much faster than Mercury's 43 arc seconds per year. They measured the rate of advance at 4.2 degrees per year, and then used Einstein's theory to calculate the combined masses of the two stars to an accuracy of six figures. They confirmed the mass measurement from the effect of gravitational redshift measurements on the arrival time of the pulses.

Even with this stunning success for general relativity, PSR 1913 + 16 had one more amazing surprise in store. For more than four years after their discovery, Hulse and Taylor continued to monitor the pulsar. According to general relativity, the pulsar and its companion are radiating gravitational waves and therefore losing energy and slowing down. The theory predicted that the orbital period should increase at a rate of 75 microseconds per year. By late 1978, they had gathered sufficient evidence to confirm the predicted effect to within an accuracy of 20%. Their achievement topped the celebrations of Einstein's centenary in 1979. By 1983, they had gathered sufficient evidence to confirm Einstein's predictions within an accuracy of 4%. This result led to Hulse and Taylor receiving the 1993 Nobel Prize for physics.

The experimental evidence in support of general relativity is now very strong. We will return to the cosmological consequences of the theory in the next chapter where we will meet what Einstein described as the biggest blunder of his career when he used his theory to predict then prevent the collapse of the Universe.

Summary

■ **The gravitational mass** of an object (i.e. its mass measured by weighing it) is equal to its **inertial mass** (i.e. its mass measured by changing its velocity)

■ **The Principle of Equivalence** asserts that the effects of gravity and of accelerated motion are identical

■ **Evidence for bending of light by gravity**
Light grazing the Sun from a star as it is occulted by the Sun during a solar eclipse
Double images of quasars due to gravitational lensing

■ **Gravitational red shift** is the decrease of the frequency of a photon escaping from the gravitational field of a massive body

■ **Black hole rules**
The Schwarzschild radius defines a sphere surrounding a black hole from which no object or light can escape
The event horizon of a black hole is the surface of the Schwarzschild sphere surrounding a black hole

Thermal radiation from a black hole due to particle-antiparticle pairs created just outside the event horizon causes a black hole to evaporate. The rate of evaporation of a black hole is insignificant, unless its mass is less than 10^{12} kg

■ **General relativity predictions**
Gravitational redshift
Deflection of light by gravity
Precession of the orbit of a planet in an elliptical orbit
Gravitational waves
Black holes

10 | THE BIG BANG

Few ideas have caught our imagination more than the idea that the Universe itself originated in a cataclysmic explosion, the Big Bang, and has continued to expand ever since. In this chapter, we look at the scientific evidence for this theory and why scientists estimate it happened about 12 000 million years ago. We also look at the conflict between the estimated age of the Universe and scientific evidence that globular clusters of stars are older than 12 000 million years. Before we meet the Big Bang, let's look at what Newton, Einstein and others thought about the Universe. We will also look at alternative theories that have been put forward to explain the Big Bang and why these theories were discarded. However, old theories in science sometimes reappear, albeit in a modified form. Science can be much more mysterious than science fiction.

Before the Big Bang theory

The Universe is thought to be expanding, causing every galaxy to move away from every other galaxy like dots on the surface of an expanding balloon. The further away a galaxy is, the faster it is receding from us. The 'Expanding Universe Theory' explains Hubble's Law and other observations. As we will see later in this chapter, the theory has led scientists to deduce that the Universe originated about 10–12 billion years ago and to make predictions about the fate of the Universe. Sir Isaac Newton, and other great scientists before Einstein, had no evidence that the Universe is expanding. Newton realized that if the Universe is finite, then the stars could not be stationary, otherwise they would all be attracted by each other's gravity into a great mass. The notion that they are rushing away from each other would prevent the collapse of a finite universe. An infinite universe would not suffer a collapse, as any star would be attracted equally in all directions by all the other stars. The stars in an infinite Universe would need to be motionless because if the distance between a star and a neighbour lessened due to their movement, they would attract each other and collide, attracting other stars to them, which would attract even more stars. An infinite Universe would be in a state of eternal collapse if the stars were not static. So Newton settled for a static and infinite Universe where the laws he discovered prevailed and where order reigned.

Newton's Universe went unchallenged for over a century, until a very simple problem was raised called **Olbers' Paradox**. This is based on the very simple observation that the night sky is dark and not bright! This seemingly trivial observation was first analysed by Heinrich Olbers in 1826. He proved mathematically that the sky would be permanently bright if the Universe consisted of an infinite number of stars. The fact that the sky is dark at night means that there is not an infinite number of stars in the Universe. Olbers originally thought that the Universe is infinite and unchanging. However, his analysis showed that this cannot be so. He reasoned that the Universe must be finite because the sky is dark. Since a finite static Universe would collapse, Olbers reckoned it must be expanding.

Einstein's error

If space is curved, a light ray could eventually return to the light source from which it was emitted, in the same way that a long-range aircraft on a journey round the world returns to its starting point. The surface of the Earth is an example of a curved two-dimensional surface which is finite and without any edge or boundary. Einstein used his general theory of relativity to predict that a finite static Universe without boundaries is possible, like the Earth's surface except in four dimensions, not two. To achieve this result, Einstein had to introduce into his equations a new type of repulsive force which acted only over a cosmological scale. This repulsive force was thought necessary by Einstein to overcome the force of attraction due to gravity which would make a finite static Universe collapse.

Einstein introduced this force into his equations by means of a 'constant of integration' which became known as the cosmological constant, represented by the symbol λ. Unlike the inverse square law of gravity,

Einstein worked out that the magnitude of the cosmological force increases with distance,

Cosmological force = $\lambda r / 3$ where r represents distance

Einstein appreciated that λ could be positive, negative or zero. A negative value would hasten the collapse of a static Universe. Einstein worked out that a zero value would also lead to collapse. A positive value would prevent the collapse. So Einstein decided that the cosmological constant was necessary and it ought to be sufficiently positive to prevent the gravitational collapse of the Universe. Just as a balloon is prevented from collapsing by its internal pressure, so a cosmological repulsive force was thought necessary by Einstein to keep the Universe static.

Einstein's model of the Universe seemed to be self-consistent with no internal contradictions. The presence of the cosmological constant, though, made fellow theoreticians uneasy, as it seemed to be an unjustified 'fiddle factor' which was not in keeping with the rigour and elegance of the Theory of General Relativity. Hubble's discovery that the distant galaxies are receding increased the unease.

The problem of the cosmological constant re-emerged in 1927 when Georges Lemaître, a Belgian priest and mathematician, discovered 'expanding Universe' solutions to Einstein's equations without the necessity of the cosmological constant. Lemaître also discovered that his solutions had been worked out five years earlier by the Russian mathematician, Alexander Friedmann, who died in 1925 before his work became known. In recognition of Friedmann's contributions, the solutions are known as Friedmann models. His most interesting model is one in which the Universe expands then contracts. Not surprisingly, the Church began to take an open-minded interest in the work of cosmologists again, several centuries after Galileo, and Lemaître became President of the Pontifical Academy of Sciences. Although Einstein described the introduction of λ as 'the biggest blunder of my life', it has been resurrected in recent decades to make the Universe older. We will return to the fate of the Universe in Chapter 13.

Constants of integration

Integration is the process of adding changes together to give the total change. A constant of integration is always part of the solution of any mathematical equation involving the rate of change of a quantity. For example, the rate of increase of velocity of an object due to gravity only just above the ground is –9.8 metres per second per second, represented by the symbol g. The minus sign is necessary if we adopt the '+ is upwards, – is downwards' convention.

■ This may be written as an equation in the form $\frac{dv}{dt} = g$,

where $g = -9.8$ m/s^2 and $\frac{d}{dt}$ means 'rate of change'.

The increase of velocity, Δv, of the object during the fall at a given time is therefore gt, where t is the flight time in seconds.

■ $\Delta v = gt$

The actual velocity at time t is the initial velocity u, plus the change of velocity Δv.

■ $v = u + gt$ where u is a constant representing the initial velocity of the object.

The initial velocity u can be positive (i.e. projected upwards), negative (i.e. projected downwards) or zero (i.e. released at rest).

The beginning of time

Scientists reckon the Universe originated between 10 000 and 15 000 million years ago in a massive explosion, the Big Bang, throwing matter and radiation outwards from the point of its creation. What preceded the Big Bang or what surrounded the point of creation are questions beyond the present realms of science as space and time were created by the expansion of the Universe and so cannot be said to have existed before the Big Bang.

Nothing can travel faster than light, so the edge of the Universe may be defined where light from the Big Bang has spread to. This is just the same as circular ripples spreading out on water as a result of dropping a stone in the water. The Universe can be imagined as an expanding sphere, like ripples spreading out on water. The furthest galaxies, formed from matter thrown outwards by the Big Bang, are thought to be moving very close to the speed of light and therefore must be very close to the edge of the expanding Universe. The distance to the furthest galaxies, using Hubble's Law, therefore provides an estimate of the distance to the edge of the

Universe. In this section, we shall look in detail at how this estimate is obtained from Hubble's Law and how it is used to determine how long ago the Big Bang occurred, which is the age of the Universe.

Using Hubble's Law to estimate the age of the Universe

The distant galaxies are thought to be receding from us and from each other at speeds approaching the speed of light. As explained in Chapter 8, Hubble's Law states that the speed of recession v of a distant galaxy increases with its distance d from us, in accordance with the equation $v = \mathbf{Hd}$ where H is the Hubble constant.

At present, this relationship is based on measuring the speed of recession of galaxies up to distances no more than about 1500 Mpc (= 5000 million light years).

The value of H is thought to lie between 50 and 100 km s^{-1} Mpc^{-1}. Figure 10.1 shows how the speed of recession increases with distance, using values for H equal to 50 km s^{-1} Mpc^{-1} for line A and 100 km s^{-1} Mpc^{-1} for line B, up to about 1500 Mpc.

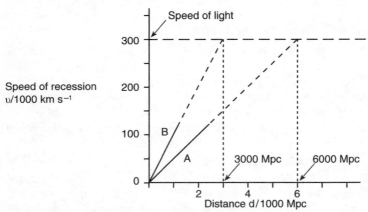

Figure 10.1   Extrapolating Hubble's Law

■ The parsec (pc) as a unit of distance was introduced on page 65. Another commonly used distance unit in astronomy, also introduced earlier in Chapter 2, is the light year, the distance travelled by light in 1 year. Since 1 parsec equals 3.26 light years, then 1 megaparsec (Mpc) is equal to 3.26 million light years.

ACTIVITY

For a galaxy at unknown distance d, its speed of recession (determined from its red shift, as explained in Chapter 8) can be substituted into Hubble's Law to calculate its distance d.

For example, suppose the speed of recession of a certain galaxy is known to be 120 000 km s^{-1}.

Rearranging $\upsilon = Hd$ gives $d = \dfrac{\upsilon}{H}$

■ Using H = 50 km s^{-1} Mpc^{-1} gives $d = \dfrac{120\,000}{50} = 2400$ Mpc ≈ 7800 million light years

■ Prove for yourself that using H = 100 km s^{-1} Mpc^{-1} gives a distance of approximately 3900 million light years.

The distance d to the galaxy therefore lies between 3900 and 7800 million light years. Clearly, a more precise value of H would give a more precise value of d.

Hubble's Law, as represented in Figure 10.1, can be used to estimate the distance travelled by light since the Big Bang. This distance is an estimate of the distance D to the edge of the Universe, since nothing can travel faster than the speed of light in free space. If D is in light years, then its numerical value is the time since the Big Bang, which is the age of the Universe.

To make this estimate, the two lines of Figure 10.1 have been extrapolated to c. The distance D to the edge of the Universe can be estimated in megaparsecs from Figure 10.1 by reading off the distance d, where the speed of recession, υ, equals c for each line. The distance can then be converted into millions of light years (Mly) by multiplying by 3.26 (see page 158), and rounding off to two significant figures.

Line A: H = 100 km s^{-1} Mpc^{-1} gives D = 3000 Mpc = 3000 × 3.26 Mly = 10 000 Mly

Line B: H = 50 km s^{-1} Mpc^{-1} gives D = 6000 Mpc = 6000 × 3.26 Mly = 20 000 Mly.

The upper value of H means that light from the edge of the Universe takes 10 000 million years to reach us. Using the lower value of H gives a time of 20 000 million years for the age of the Universe.

The link between the Hubble constant and the age of the Universe

Some more values of H, and the corresponding age estimates given by the above method, are presented in Table 10.1.

Table 10.1					
The Hubble constant, H, in km s^{-1} Mpc^{-1}	40	50	80	100	200
Age of the Universe in millions of years	25 000	20 000	12 500	10 000	5000

The table shows that the age of the Universe can be calculated in millions of years by dividing H into 1 million.

$$\text{Age (in millions of years)} = \frac{1\,000\,000}{H \text{ (in km s}^{-1} \text{ Mpc}^{-1})}$$

If H is converted into Myr^{-1} as explained below, the above equation becomes

$$\text{Age (in millions of years)} = \frac{1}{H}$$

The unit of H

■ The speed of light, c = 300 000 kilometres per second therefore 1 million light years = 300 000 T_M kilometres, where T_M is the number of seconds in 1 million years.

■ 1 Mpc = 3.26 million light years = 3.26 × 300 000 T_M kilometres ≈ 1 000 000 T_M kilometres.

Hence 1000 km s^{-1} per Mpc = $\dfrac{1000 \text{ kilometres per second}}{1\,000\,000\ T_M \text{ kilometres}}$

= $\dfrac{1}{1000\ T_M}$ per second   = $\dfrac{1}{1000}$ per million years

Therefore 1000 km s^{-1} per Mpc = $\dfrac{1}{1000}$ per million years = 0.001 Myr^{-1}

■ To see how this conversion operates, consider H = 100 km s^{-1} per Mpc. Using the above conversion gives H = 0.0001 Myr^{-1}.

Therefore $\dfrac{1}{H} = \dfrac{1}{0.0001 \text{ Myr}^{-1}}$ = 10 000 Myr

The Big Bang explanation of Hubble's Law

Consider two points separated by distance d moving apart at a steady rate. Suppose the two points were initially at the same position, so that at time t later, their separation $d = \upsilon t$, where υ is the speed of separation. We can therefore write $\upsilon = \dfrac{1}{t} d$ or $\upsilon = Hd$ if we let $\dfrac{1}{t} = H$.

Is the Hubble constant a universal constant? If the Hubble constant differed in the past, the age of the Universe would be different to the value estimated opposite. A reduced value of H in the past would make the age of the Universe greater. For example, if H was equal to 50 km s^{-1} for most of the lifetime of the Universe except the last 2 billion years, the age of the Universe would be 22 billion years (= 20 billion years for H = 50 km s^{-1} per Mpc plus the last 2 billion years). We will look again at H in Chapter 13 when we consider a more sophisticated explanation of Hubble's Law.

ACTIVITY

Make a model of the expanding Universe

Mark a balloon with dots and inflate the balloon gradually. The dots move apart as the balloon is inflated. There is no central point on the surface of the balloon and all dots are equivalent to each other.

- The distance from any one dot to any other dot increases with time.
- The further away two dots are from each other on the surface, the faster they move away from each other.

Half-inflate the balloon and draw the shortest possible line between two points, marking the line into equal intervals. Repeat the procedure for two more dots at different spacing, using the same interval length as before. The distance d between each pair of points is equal to the number of intervals × length of each interval. Now inflate the balloon and you will see that each interval becomes longer, but the number of intervals doesn't change between each pair of points. Which pair of dots separates fastest? The answer is the pair that are further apart – because there are more intervals separating this pair, and each interval lengthens at the same rate.

The history of the Hubble constant

Edwin Hubble established the velocity–distance relationship for galaxies in 1929, making measurements on about two dozen galaxies at distances up to 2 Mpc away. His original value of the Hubble constant was 530 km s^{-1} Mpc^{-1} which produces an estimated age of the Universe of little more than about 2000 million years. This estimate conflicted with the age of the Earth at 5000 million years, determined by a radioactive dating method involving the measurement of the proportion of uranium to lead in uranium ore. Uranium nuclei are unstable and disintegrate to form lead nuclei in a process with an effective half-life of about 4500 million years. A sample of uranium ore containing 50% uranium and 50% lead would be 4500 million years old, assuming it was pure uranium at the time of its formation. With this assumption, by measuring the relative proportion of uranium to lead, the age of the sample can be determined accurately.

Hubble's estimate of the Hubble constant was based on distance estimates which had a high degree of uncertainty. By 1936, improved measurements halved the Hubble constant, doubling the age of the Universe to more than 4000 million years – still not enough to make the Universe older than the Earth. The expanding Universe model was called into question even though Hubble's Law was established scientific fact up to about 2 Mpc. The difficulty was made much worse when astronomers discovered that clusters of stars known as globular clusters in the Halo of the Milky Way are as much as 15 000 million years old.

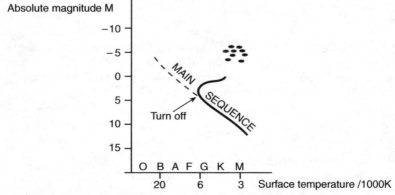

Figure 10.2   HR diagram for a globular cluster

The story of globular clusters

Globular clusters are observed in a broad part of the sky, which is towards the centre of the Milky Way galaxy, above and below the galactic plane. Each cluster might contain over a million stars, held together by their own gravity. The distance to a globular cluster can be determined by timing the period of one or more Cepheid variable stars in it, as explained on page 110. The absolute magnitude of the other stars in the cluster can then be worked out and plotted on a Hertzsprung–Russell diagram. In general, the stars in a globular cluster lie along the Main Sequence up to a point where they leave the Main Sequence for the red giant branch. As explained on page 89, the mass of the stars on the Main Sequence is known to increase from about 0.1 solar masses to about 30 solar masses at most. The point where the stars of a globular cluster turn off the Main Sequence can therefore be used to determine the mass of a star in the cluster at the end of its Main Sequence life. Since each kilogram of hydrogen in a star is known to release 7×10^{14} J (see page 105), and it is known that a star at this stage has used 20% of its hydrogen through fusion to helium, the age of the heaviest star can be worked out from its mass and its luminosity (i.e. the energy it radiates away each second):

$$\text{age in seconds} = \frac{0.2 \times \text{mass in kilograms} \times 7 \times 10^{14}}{\text{luminosity in joules per second}}$$

For example, suppose the stars in a globular cluster turn off the Main Sequence at about 1 magnitude higher than the Sun, corresponding to a luminosity of about 1×10^{27} J/s and a mass of about 2 solar masses ($= 4 \times 10^{30}$ kg). The uncertainties in the measurements give a Main Sequence life for the oldest stars in such a cluster between 12 000 and 18 000 million years. Clearly, this estimate presented astronomers with a major problem – stars older than the Universe!

Does gravity affect Hubble's Law?

Gravity is an attractive force that acts between any two objects. The force of gravity between two objects becomes less and less with increased separation but it does not become zero. Gravity has an infinite range, and therefore the force of gravity between receding galaxies acts against their motion.

Using Newton's laws, as explained in Appendix 3, it can be shown that the effect of gravity is to slow the expansion of the Universe down in accordance with the equation

$$\upsilon = \frac{2}{3t}\, d \quad \text{instead of } \upsilon = \frac{d}{t}$$

The Hubble constant H is therefore $\frac{2}{3t}$ according to this model, giving an estimated age for the Universe equal to $\frac{2}{3H}$ instead of $\frac{1}{H}$

Regardless of the details of how the effect of gravity is worked out, clearly the overall effect is to make the estimated age of the Universe smaller than 1/H. The discrepancy between the age of globular clusters and the age of the Universe is even worse when gravity is taken into account. This increased the need to find out if Hubble's Law is generally valid further away and, if so, to obtain a more reliable estimate for H. In addition, evidence from further away would give an indication if H was the same in the past as it is now. We will look at the consequences of a changing value of H in Chapter 13.

Looking deeper and deeper into space

To check Hubble's Law is true at great distances, astronomers need to be able to observe galaxies further and further away in as much detail as possible. The detail that can be seen with a telescope depends on how wide the telescope mirror is. The wider the mirror, the more detail that can be seen in an extended image because diffraction of light by the mirror is reduced.

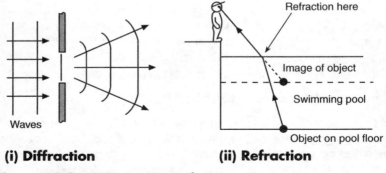

(i) Diffraction   **(ii) Refraction**

Figure 10.3   Wave properties

Diffraction is the spreading of waves whenever they pass through a gap. The narrower the gap, the greater the diffraction; conversely, the wider the gap, the less the amount of diffraction. Diffraction affects image formation by lenses and mirrors. Each point of an image formed by an optical system is due to light from the corresponding point of the object being focused by the lenses or mirrors of the system. Diffraction smears out the points of the image, reducing the amount of detail in the image.

Images seen through ground-based telescopes are smeared out due to atmospheric refraction, as well as due to diffraction by the objective mirror. This is why stars twinkle. Refraction is the bending of light when it passes from one transparent medium to another. For example, a swimming pool appears shallower than it really is because a light ray from the bottom of the pool bends towards the surface as it emerges into air. If the water surface is smooth, anyone at the pool side sees an image of the bottom of the pool 'higher up' than it really is. Ripples on the surface break the image up and make it difficult to see. Light passing through the Earth's atmosphere is affected by 'ripples' in the atmosphere caused by local temperature variations. The image of a star moves about slightly because of this effect. Beyond a diameter of about 10 cm, atmospheric refraction limits the detail seen through a ground-based telescope. This is why the Hubble Space Telescope, in orbit above the atmosphere, is able to see much more detail than a ground-based telescope. In addition, it is unaffected by atmospheric absorption of light, so it collects more light than a similar telescope on the ground.

ACTIVITY

Use a pair of binoculars, or a telescope, to observe a binary star such as Mizar in Ursa Major. The star map in Figure 2.2 should help you to find Mizar. You should be able to see its binary companion Alcor when you look through binoculars or a telescope. The entrance to the telescope is much wider than your eye pupil so the light is diffracted much less than when you look with the unaided eye. Consequently, the images are distinct when you view through binoculars or a telescope, but not when you view unaided.

Hubble's progress

In 1949, the 200 inch telescope at Mount Palomar was completed and used by Hubble for the first time. This high-altitude telescope was able to distinguish individual stars much better than any existing telescope. As a result, more accurate measurements were made, reducing the Hubble constant to about one third of the first estimate. By 1958, further detailed measurements on many more galaxies, up to distances of about 20 Mpc, reduced the Hubble constant to about 75 km s^{-1} Mpc^{-1}, giving an estimated age for the Universe of about 9000 million years (= 2/3H). However, the distance measurements were subject to large inaccuracies and before the Hubble telescope was first used, the accepted value of the Hubble constant lay between 50 and 100 km s^{-1} Mpc^{-1}, corresponding to an estimated age of the Universe between about 8000 and 15 000 million years.

The Hubble Space Telescope (HST)

Since its 2.4 m mirror was corrected in 1993, the Hubble Space Telescope has captured stunning images of many objects in space. These images are much much clearer and brighter than from existing ground-based telescopes. Above the Earth's atmosphere, the images cannot be smeared out and distorted by the Earth's atmosphere.

A large telescope has a wide concave mirror as its objective, reflecting as much light as possible into the eye piece. Figure 10.4 shows how this is achieved.

The wider the mirror, the more light can be reflected into the eye piece. Increasing the diameter of a mirror by a factor of 10 enables 100 times as

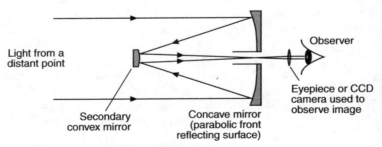

Figure 10.4   A reflector telescope

much light to be collected because the collecting area is proportional to the square of the mirror's diameter. This corresponds to an extra 5 magnitudes in terms of the brightness of a star since, by definition as detailed on page 68, a hundredfold increase in light intensity corresponds to a change of 5 magnitudes.

The wider the mirror is, the more detail that can be seen in an extended image. This is because diffraction is reduced, as explained on page 165.

Also, as explained on page 165, images seen through ground-based telescopes are smeared out due to atmospheric refraction, as well as due to diffraction by the objective mirror. In orbit above the atmosphere, HST is able to see much more detail than a ground-based telescope. In addition, because it is unaffected by atmospheric absorption of light, it collects more light than a similar telescope on the ground.

A task for the Hubble Space Telescope

Using HST, astronomers have been able to determine the distances to stars and galaxies more accurately. As explained on page 65, the distance to a nearby star is measured using the parallax method which involves measuring its position against the background of distant stars at six-month intervals. Because the Earth moves round the Sun by $180°$ in this time, the nearby star appears to shift its position, as viewed from the Earth. The absolute magnitude M of the star can then be calculated from its apparent magnitude m using the conversion graph in Figure 6.2, which is represented in this section as Table 10.2. The discovery that the time period of a Cepheid variable star depends on its absolute magnitude gave astronomers a means of measuring the distances to Cepheid variables beyond the range of the parallax method simply by timing the period and using the link between M and the time period to determine M.

As explained on page 70, absolute magnitude M is the magnitude a star would have if it was at a distance of 10 parsecs. Table 10.2 on the next page shows the relationship between the absolute magnitude M and the apparent magnitude, m, of a star. For example, a star at a distance of 100 pc with an apparent magnitude, m, of +6 has an absolute magnitude M = +1.

Table 10.2							
d/pc	0.01	0.1	1	10	100	1000	10 000
M – m	15	10	5	0	–5	–10	–15

One of the first experiments carried out using HST involved observing individual Cepheid variables in M100, a spiral galaxy with a known red shift which gave its speed of recession at 1400 km s^{-1}.

Using the telescope, the variability in brightness of Cepheid variables too faint to detect using ground-based telescopes was measured. The distance to M100 was thus measured at 17 Mpc, giving a value of the Hubble constant of 80 km/s Mpc^{-1}, accurate to 20%. This gives a value for the age of the Universe of about 8000 million years. Subsequent measurements in 1995 gave a value for the Hubble constant of 57 km/s Mpc^{-1}, corresponding to an age of 11 000 million years. The range of uncertainty in the measurements does give an upper limit for the age of the Universe which is above the lower limit of the estimated age of the oldest globular cluster. More accurate measurements could well resolve the issue of the age of the Universe by proving that the oldest globular clusters are younger than the age of the Universe. If not, cosmologists have a considerable amount of rethinking to do.

At the time of writing, work continues using HST to measure the distances to supernovae 100 times further away than the faintest Cepheid variable star. This will enable the Hubble constant to be determined accurately out to much greater distances than at present, instead of being extrapolated. The results thus far give a value of 65 km s^{-1} Mpc^{-1} although at the furthermost distances of the order of 1500 Mpc, there is some evidence for a slight increase. Perhaps this indicates accelerated expansion of the Universe although many more measurements need to be made to obtain firm evidence. A value of 65 km s^{-1} Mpc^{-1} for H gives an age of about 10 000 million years.

Evidence for the Big Bang theory

If the number of stars is not infinite, gravitational attraction between galaxies would cause the Universe to collapse unless either it is expanding as Olbers deduced (in which case gravity would reduce or eventually stop the rate of expansion) or else matter must be created continuously to prevent it collapsing. This latter theory, known as the **Steady State theory**

of the Universe, was supported by many prominent scientists until recently, in spite of Hubble's Law. In 1928, the famous astronomer Sir James Jeans spoke of 'points at the centres of nebulae where matter is continually poured into our Universe from some other dimension' and this view was still maintained by many eminent scientists until the 1960s. The Steady State theory was developed from Jeans' ideas by Hermann Bondi, Thomas Gold and Fred Hoyle in 1948. The theory assumed that the Universe on a sufficiently large scale is the same at all locations and at all times. Assuming the Hubble constant does not change, the theory required the continuous creation of matter to make up for the spreading out of matter due to expansion. Points of creation referred to as white holes were envisaged, the opposite to black holes.

The reason for their support of the Steady State theory lay in the discrepancy between the age of stars in globular clusters and the age of the Universe as estimated from the Hubble constant. The present value of the Hubble constant is about 65 km s^{-1} Mpc^{-1} (where 1 km s^{-1} Mpc^{-1} = 3.2×10^{-20} s^{-1}). Thus 1/H gives an age for the Universe of 4.8×10^{17}s which is about 15 billion years. Gravity slows the expansion down and gives 2/3H as the age of the Universe which is about 10 billion years, using the present estimate of H. At the time the Steady State theory was put forward, the Hubble constant was thought to imply an age of little more than 4 billion years for the Universe, assuming the Big Bang theory. Improved measurements reduced the value of the Hubble constant to less than about 100 km s^{-1} Mpc^{-1}, corresponding to an age for the Universe of about 7 billion years, younger than the age of the oldest stars, reckoned to be about between 12 and 15 billion years. Clearly those who argued in favour of the Steady State theory were distinctly unimpressed by a Universe younger than its own stars! However, their difficulties in convincing the Big Bang theorists deepened when the results and conclusions from an extensive survey of radio sources were published in 1955 by Martin Ryle at Cambridge. Using a high-resolution radio telescope, Ryle proved that the concentration of radio sources increased at large redshifts. This finding weakened the basic principle of uniformity in the Steady State theory, the idea that the Universe should be the same on a large scale everywhere. Further surveys over the next decade supported Ryle's discovery. According to Big Bang theory, the red shift of a distant galaxy is a measure of its distance. The concentration of radio sources at high redshifts seemed to indicate an excess of radio galaxies at some time in the distant past.

The dispute between the two theories was settled unambiguously in favour of the Big Bang by two major discoveries in the following decade. The Big Bang theory is now supported by very strong evidence based on three discoveries.

1 Microwave background radiation from all directions in space was first detected by Arno Penzias and Robert Wilson in 1965. They converted a satellite receiver system into a detector for radio astronomy by fitting a large reflector horn to the system. When they came to test the modified system, they noticed background radiation was detected whichever direction the horn was turned. The background radiation was the same in all directions. Even though their receiver horn was able to distinguish radio sources as close as one sixtieth of a degree, they found the radiation was the same right round the sky. The discovery was taken up by other scientists and within a few months, the spectrum of the radiation had been measured and found to be the same as the thermal radiation spectrum of a body at a temperature of 2.9 degrees above absolute zero.

The presence of microwave background radiation is inexplicable in terms of the Steady State theory of the Universe. This theory gives no reason for its existence. However, the Big Bang theory does provide a natural explanation, as the radiation is thought to be radiation released in the Big Bang that became longer and longer in wavelength as the Universe expanded. The energy, and hence the frequency of the photons of radiation, released in the Big Bang would have been very high, corresponding to a temperature of billions of degrees. According to the Big Bang theory, as the Universe expanded, it became cooler and the radiation was stretched out more and more as the Universe expanded. The wavelength of the radiation therefore became longer and longer, and is now about 1 mm at peak intensity.

In 1989, the Cosmic Background Explorer satellite (COBE), launched into orbit round the Earth to map out this background radiation, discovered slight variations in it. We will return to the significance of this slight unevenness in Chapter 12.

2 Stars and galaxies contain about three times as much hydrogen as helium. This observation can be explained as a consequence of the cooling of the Universe after the Big Bang. Above a temperature of 10 000 million degrees (= 10^{10}K), neutrons and protons break free from nuclei. As the Universe cooled, neutrons and protons joined to form helium nuclei at

10^{10}K, leaving excess protons as hydrogen nuclei. Each helium nucleus locked up 2 protons and 2 neutrons, leaving 12 excess protons as hydrogen nuclei, corresponding to a hydrogen:helium mass ratio of three to one. Hydrogen is three times as abundant as helium because there were 14 protons for every 2 neutrons before the nuclei of atoms were formed.

Why should there have been 7 times as many protons as neutrons when the early Universe was at a temperature of 10^{10}K? The answer to this question lies in the fact that the neutron is slightly heavier than the proton, and therefore the energy associated with its rest mass is slightly larger than that for the proton. We will see in Chapter 11 that the proton and the neutron are made up of smaller particles called **quarks**. More energy is needed to form a neutron than a proton, so the formation of a proton is more likely than the formation of a neutron. This energy difference is about the same as the thermal energy of a proton or a neutron at a temperature of about 10^{10}K, and is just enough to account statistically for the fact that there is approximately one neutron for every seven protons.

The exact temperature needed to give the one-to-seven ratio is 3×10^9K. As this is the temperature at which nuclei would disintegrate by collision, if matter was heated and heated, the Big Bang theory makes it possible to explain why the neutron-to-proton ratio is 1:7, and hence why there is about three times as much hydrogen in the Universe as helium.

The actual temperature of 3×10^9K is a billion times greater than the temperature of the microwave background, suggesting that photon energies were a billion times greater, and photon wavelengths were a billion times shorter, when the formation of protons and neutrons occurred. We shall return to this point in Chapter 12 when we look back at the history of the Universe.

3 Explanation of Hubble's Law: the Big Bang theory of the origin of the Universe leads to the expanding model of the Universe, which provides an explanation of Hubble's Law. The Steady State theory of the Universe requires the continuous creation of matter to explain Hubble's Law. There is no evidence for the continuous creation of matter. Also, the Steady State theory cannot explain the microwave background or the hydrogen-to-helium ratio.

Summary

■ **Olbers' Paradox**: the Universe is not infinite as the sky would be permanently bright if it was infinite

Models of the Universe

■ **Newton's universe** is static and infinite
■ **Einstein's static universe** requires the introduction of a cosmological force of repulsion
■ **Friedmann's universe** expands without the necessity of a cosmological force
■ **The Steady State model** supposes that matter is continuously created as matter spreads out
■ **The Big Bang model** supposes that the Universe was created in an explosion at a point billions of years ago, and has been expanding ever since
■ **Hubble's Law**
The speed of recession of a distant galaxy = Hd, where d is its distance, and H is the Hubble constant.
The value of H is thought to be about 65 km/s Mpc^{-1} or 65/1 000 000 Myr^{-1}

■ **Age of the Universe** (in Myr) = $\dfrac{1}{H}$ where H is in Myr^{-1} neglecting gravity

$$= \frac{2}{3H} \text{ taking account of gravity}$$

■ **Evidence for the Big Bang theory**
Hubble's Law
Microwave background radiation
Three-to-one ratio of hydrogen to helium

11 | FUNDAMENTAL FORCES

Astronomers observe and study distant galaxies rushing away from us at speeds approaching the speed of light. The further away a galaxy is, the longer its light has taken to reach us. The most distant galaxies were formed billions of years ago. To observe them is to look back in time when the Universe was much younger, much hotter and much smaller. The study of particles of matter on the smallest possible scale under high-energy conditions now preoccupies many scientists in every continent in a long-running attempt to understand how particles interact, and why they have the properties that they have. This field of study relates closely to the work of astronomers and cosmologists attempting to understand how matter formed from radiation in the Big Bang, and why matter formed into stars and galaxies. Science on the largest possible scale meets science on the smallest scale as cosmologists search for the details of the Big Bang. In this chapter, we will look at how and why particles interact and why the knowledge gained by studying these interactions has cosmological consequences.

Forces and fields

Forces make objects move or change the shape of objects. How many different ways can you think of to make an object move? An object moves when it is released above the floor because it is attracted towards the Earth by the force of the Earth's gravity. A magnet can pick up paper-clips because it attracts them with a magnetic force. A plastic ruler rubbed with a dry cloth can pick up bits of paper because it attracts them due to static electricity. It may come as a surprise to learn that there are just two types of force at work in all the above situations, one of which is gravity. All the other situations are where the force between charged particles is at work.

The force of gravity between any two objects is bigger the larger the mass of each object. Gravity acts between two objects because the objects possess mass. Exactly how the force of gravity acts between them is not known, even now almost a century after Einstein published the General Theory of Relativity. Precisely what gives an object its mass is another unanswered question at present. We saw in Chapter 9 that gravity makes space-time curved. But what causes an object to possess mass, and why does the mass of an object increase if it is supplied with energy? Gravity is a familiar force, yet it raises so many fundamental questions.

In contrast to the mysteries of gravity, the force between charged particles is less familiar, yet more is known about this force. In 1861, James Clerk Maxwell showed mathematically that electricity and magnetism are one and the same. A century before Maxwell, the law of force between two charged particles had been worked out by the French scientist, Charles Coulomb. He proved by experiment that the force between two charged particles varied with distance according to the inverse square law – just as the force of gravity between two masses varies in accordance with Newton's Law of Gravitation. Double the distance and the force becomes one quarter as strong: treble the distance and it becomes one ninth as strong.

Current electricity from batteries was shown to be due to the flow of electric charge round an electric circuit. Scientists soon discovered that current electricity had several useful properties, including its heating effect, its chemical effect now used in connection with metal refining, and its magnetic effect. This latter effect led to the invention of the electromagnet, in which the magnetic effect of an electric current is used to magnetize iron. In 1829, Michael Faraday discovered how to generate current electricity by making a magnet move in and out of a coil of wire which was part of a circuit. Faraday's discovery was arguably one of the greatest discoveries of the century, for it meant that electricity could be generated by dynamos and then distributed to homes, offices and workshops by a grid of cables. The electricity network that supplies every home, every place of work and the transport system was developed gradually in the nineteenth century following Faraday's great discovery. Candles or batteries for lighting, coal for heating and for steam power to industry and transport have all been replaced by electrical power from generator stations. It is said that when Faraday demonstrated his discovery to an audience at the Royal Institution in London, he was asked by a member of the audience what was the use of his discovery. In a memorable reply, he asked 'What use is a new baby?'

Michael Faraday 1791–1867

Faraday was a most unusual scientist. He gained an appetite for science in his first job as a bookbinder's apprentice. As well as binding the books, he read them avidly, and learned about electricity from books such as Encyclopedia Britannica. He wrote to Sir Humphrey Davy at the Royal Institution in London and gained a position as an assistant as a result. His research into electromagnetism led to his famous discovery of how to generate electricity. He developed great insight into the link between electricity and magnetism, which led to many important applications including the transformer, the electric motor and the electric telegraph.

Faraday's ideas were taken up by Maxwell, who was a mathematical physicist rather than an experimenter. Maxwell showed that an alternating current in a wire generates waves of electric and magnetic fields that spread out from the wire into the surroundings. He proved that such electromagnetic waves travel through air at a speed of 300 000 km/s, at the speed of light. This left no doubt that light itself consists of electromagnetic waves, and that other invisible forms of electromagnetic waves must exist. Within a few decades, the properties of the complete spectrum of electromagnetic waves had been discovered, investigated and set to work in a vast range of applications from long-distance radio communications to X-rays used in hospitals. At a fundamental level, Maxwell proved that all magnetic effects are manifestations of the force between charged particles. In effect, Maxwell showed that magnetic and electric forces are different aspects of the same 'electromagnetic' force and are governed by a single law, namely Coulomb's law of force between point charges.

The everyday force

Most everyday situations involving force are either due to gravity or due to the electromagnetic force. Take out gravity and you are left with just the electromagnetic force. This is the force responsible for keeping electrons in atoms, attracted by the electrical force to the nucleus. Atoms become charged or 'ionized' as a result of gaining or losing electrons. Almost all chemical and biological effects are due to atoms, ions or electrons interacting with each other through the electromagnetic force. Solids gain their strength, and liquids don't disperse into the air because the atoms are held together by electromagnetic forces. Friction is because the atoms and electrons on two surfaces in contact attract each other. Everything you do that doesn't involve gravity is because of the electromagnetic force. However, gravity and the electromagnetic forces are not quite enough to explain your existence and the existence of all the objects around you and in most of the rest of the Universe. Some other force must exist to prevent the protons and neutrons in each atomic nucleus from flying apart. This force is called the **strong nuclear force**. It doesn't stretch much beyond the nucleus, unlike gravity and the electromagnetic force which stretch to infinity. If the strong nuclear force extended to infinity too, the Universe would probably have never expanded beyond a gigantic ball of neutrons and protons. We will return to the strong nuclear force later in this chapter.

Force fields

The interaction between two objects due to gravity can be thought of in terms of gravitational fields created by each object. One object in the force field of another object experiences a force towards the object that is the source of the field. The effect is mutual, so the two objects exert equal and opposite forces on each other. The same idea can be used to describe how two charged objects interact electrically. The charge on each object creates an electromagnetic field surrounding the object. Any other charged body in that field experiences an electromagnetic force due to the field. Again, the effect is mutual, so the two objects exert equal and opposite forces on each other.

The concept of a field is a *description* of how objects interact. The electromagnetic waves that spread out from a TV transmitter aerial make electrons oscillate in distant receiver aerials, enabling TV signals to be picked up so you can watch your favourite programmes. As we saw on page 51 in Chapter 5, electromagnetic waves are quantized as photons. Each photon is a packet of electromagnetic waves that carries well-defined energy in proportion to the wave frequency. In the transmitter aerial, electrons are forced to oscillate which makes them emit photons. The electrons in the receiver aerial absorb some of the photons from the transmitter, causing the electrons in the receiver to oscillate. A photon absorbed by an electron in the receiver represents an interaction between that electron and an electron in the transmitter. This idea was perhaps behind the mathematical theory of electromagnetic interaction developed by Richard Feynman in 1948. Feynman proved that electrons interact by exchanging **virtual photons**. Unlike a real photon which carries energy away from an oscillating electron, a virtual photon can be exchanged regardless of the motion of the two electrons involved. Predictions from Feynman's theory were subsequently confirmed experimentally, leaving little room for doubt that charges interact through the exchange of virtual photons. Feynman translated his complicated mathematical formulas into diagrams which became known as Feynman diagrams.

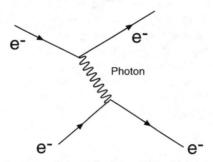

Figure 11.1 A Feynman diagram representing the interaction between two electrons

Richard Feynman 1918–1988

In his schooldays in the 1930s when he developed a reputation for fixing radios, Richard Feynman proved he was no ordinary student. By combining his immense mathematical talent with powerful insight and intuition in physics, he became one of the foremost physicists of the 20th century. His books and lectures inspired several generations of physics students in every continent. After the Challenger space shuttle disaster, he proved the cause of the disaster in the subsequent investigation by means of a simple demonstration of what went wrong.

Particles in the nucleus

Every atom contains a nucleus which consists of protons and neutrons. The size of the nucleus in relation to the atom is about the same as a marble to a large football stadium. The marble at the centre of the stadium is like the nucleus at the centre of the atom. The electrons in the atom are in allowed orbitals around the nucleus, defining by their presence the rest of the atom. The electrons in an atom carry a negative charge, and are held in the atom by the electrical attraction of the nucleus which carries a positive charge. Opposite charges attract, so each electron in its orbital experiences a force towards the nucleus which keeps the electron in its orbital and prevents it from leaving the atom.

Protons are positively charged and neutrons are uncharged. An obvious question to ask is what holds the nucleus together when its protons must repel each other because they carry like charges and its neutrons are uncharged. The fact that the nuclei of most atoms are stable tells us that there must be an attractive force holding the protons and neutrons of a nucleus together. Also, this force must be stronger than the electrical force of repulsion between the protons. This is the role of the strong nuclear force. It acts equally as an attractive force between protons and neutrons, and has a range of no more than about 10^{-15} metres, the size of the nucleus of a light atom.

The lightest nucleus is the hydrogen nucleus which is a single proton. The next lightest stable nucleus is the helium nucleus, which consists of two protons and two neutrons. When hydrogen is fused to form helium inside a star, protons are forced together against their electrical repulsion until they are so close that the strong nuclear force attracts them together. When this process occurs, four protons are fused in stages to form a helium nucleus which consists of two protons and two neutrons. Two of the four protons turn into neutrons in a process which is governed by an interaction known as the **weak nuclear force**. The strong nuclear force holds the protons and neutrons together in the nucleus. However, the particles in the nucleus can swap identity if there are too many of one type crowded together.

Mystery particles from the nucleus

We saw in Chapter 5 that a photon of gamma radiation can suddenly turn into a particle and its antiparticle in a process known as pair production. The energy of the photon must be sufficient to create the mass of the particle and antiparticle in accordance with Einstein's famous equation, $E = mc^2$. The inverse process, known as annihilation, happens when a particle meets its antiparticle to release gamma radiation. Again, the process must obey Einstein's equation as the total energy (including mass converted to or from energy) must be the same after the change as before.

The principle that energy is always conserved when it changes from one form into other forms was established in 1840 by the English scientist James Joule, who conducted very precise heat experiments to test the principle. Joule's experiments led to the acceptance of the principle as a fundamental law of science. However, as with any scientific law, experiments can do no more than support a law or else disprove it. Agreement between a theory and an experiment enhances the theory but can never prove it. The principle of conservation of energy was seriously challenged for some years by the experimental finding that energy seemed to disappear in the radioactive process of beta decay (see page 58). Measurement of the energy spectrum of the beta particles from any radioactive source showed that there is a spread of speeds of the emitted particles up to a maximum. The maximum speed was shown to correspond to the energy released by the change, which can be worked out using $E = mc^2$. The puzzle was why do the beta particles not all carry away the same amount of energy? What do becomes of the energy that is released and not carried away by a beta particle? In the 1930s, the Italian scientist Enrico Fermi put forward the theory that another particle was also emitted at the same time as the beta particle was emitted. He called this other particle the **neutrino** or the 'little neutral one'. He reasoned it must be neutral like the neutron, otherwise it could be detected from its charge. He also worked out that it must have almost zero rest mass as it could not be detected at the time. Fermi's neutrino theory was confirmed in 1956 when two American scientists, Reines and Cowan, successfully detected neutrino reactions in a tank of fluid constructed for the experiment, next to a nuclear reactor. They reckoned that a neutrino from the reactor could interact with a proton to convert the proton into a neutron and produce a positron at the same time. The positron, as the antimatter particle of the electron, would travel no more than a fraction of a millimetre from its point of production before being annihilated by an electron. Two gamma photons would be released in opposite directions in this process, so Reines and Cowan sought evidence that detectors exactly opposite each other on either side of the tank would be triggered simultaneously. They detected such events at a rate of about three per hour over a running period of almost 1400 hours when the reactor was operating. They worked out from their measurements that neutrinos and their antimatter counterparts are over a million million million times less likely to interact with a proton than a neutron is.

A neutrino and a positron are released when a proton changes into a neutron. Such changes occur in vast numbers inside the core of every star, including the Sun, as a result of nuclear fusion when hydrogen is converted into helium. Huge numbers of neutrinos stream out of the Sun every second, travelling at or near the speed of light, passing through any matter in their path with scarcely any effect. At this very moment, billions of neutrinos are passing through your body from the Sun with no observable effect. Richard Feynman jokingly likened neutrinos to teenagers 'passing through the house almost at the speed of light without interacting'. In 1987, when the supernova SN1987A was detected, astronomers successfully alerted their physicist colleagues working on neutrino research to check their detectors to confirm that extra neutrinos were being detected. When they looked at the recordings from their detectors, they found that the neutrino readings had gone up hours before the supernova was observed. They had detected the supernova first because the neutrinos had been able to race through the collapsing star much faster than photons could, enabling the neutrinos to speed ahead through space. As we shall see in Chapter 13, these elusive mysterious particles are now of great interest to astronomers and cosmologists.

Space invaders

The neutron was discovered in 1932 by James Chadwick at Cambridge. He found out how to accelerate alpha particles to high speeds and used them to bombard thin metal foils. He discovered that the foil emitted a very penetrating form of radiation which knocked protons out of the atoms in a wax block. From his measurements, he worked out that the protons were knocked out by uncharged particles of about the same mass. He realized these uncharged particles fitted the description of the neutron, the particle suggested by Lord Rutherford to account for the mismatch between the number of particles in the nucleus and the number of protons in it.

Chadwick's discovery of the neutron completed a tidy picture of the atom with its nucleus consisting of neutrons and protons surrounded by electrons. However, even at the time of Chadwick's discovery, there were signs that matter at this fundamental level is more complicated. The positron was discovered about this time, and a particle that became known as a meson (because it is heavier than the electron but lighter than a proton) was discovered in cloud chamber photographs. These particles, now known as muons or mu-mesons, were known to originate at high altitude and were reckoned to be the result of high-energy particles referred to as 'cosmic rays' from space crashing into the Earth's atmosphere. Patrick Blackett was awarded the Nobel Prize for physics in 1948 for his pre-war work at Cambridge where he had developed a detector-controlled photographic system for taking cloud chamber photographs only when a shower of cosmic rays was detected. Blackett introduced the term 'cosmic rays' to describe the showers of particles captured on film in such events. After the Second World War, more new particles were discovered from cloud chamber photographs and from photographic plates directly exposed to cosmic radiation at high altitude. Particles called pions or pi-mesons were discovered and recognized as the carriers of the strong nuclear force. These particles were predicted by the Japanese physicist Hideki Yukawa. He reckoned that neutrons and protons in the nucleus attract each other by exchanging lighter particles. These were thought to be muons at first, but subsequent tests showed that the properties of muons did not fit the properties of Yukawa's exchange particles. Pions were discovered by Cecil Powell in 1947 at Bristol from the tracks created by cosmic ray collisions with nuclei in photographic emulsions. Powell showed that the properties of pions matched Yukawa's exchange particles exactly and was awarded the 1950 Nobel Prize for physics for his work.

Smaller and smaller

High-energy accelerators were developed after 1945 to study the collisions between nuclei and fast-moving particles. These accelerators used high-frequency alternating voltages to accelerate particles along evacuated tubes to hit their targets. Experiments using these high-energy accelerators led to the discovery of hundreds of different short-lived particles created in the debris of the collisions. The properties of these particles were measured in the search to find an underlying pattern. More new particles were discovered each time a more powerful accelerator was built. The simple picture of the atom, with its nucleus consisting of protons and neutrons surrounded by electrons, no longer provided an explanation of the myriad host of particles created in these experiments. Order began to emerge when the Stanford Linear Accelerator in California was built to accelerate electrons through voltages up to 50 000 million volts. At such high energies, the mass of an electron becomes many times greater than its rest mass and much larger than the rest mass of a proton or a neutron. An electron at this speed is therefore capable of penetrating deep into a nucleus. By studying the scattering of electrons in such experiments, scientists realized that every proton and every neutron consists of three smaller particles. These particles were dubbed **quarks** by American theoretical physicist, Murray Gell-Mann, from the line 'Three quarks for Muster Mark' written by James Joyce in *Finnegan's Wake*. Gell-Mann first put forward the idea of quarks to explain the multiplicity of particles discovered from earlier cosmic ray and accelerator experiments. Gell-Mann's theory was based on the idea that:

a proton contains two 'up' quarks and a 'down' quark
a neutron contains two 'down' quarks and an 'up' quark
a third type of quark, **the 'strange' quark**, is very short-lived and decays to one of the other two types of quark.

Gell-Mann's theory was confirmed by the Stanford experimenters and Gell-Mann was awarded the 1969 Nobel Prize for physics.

Quark rules

A proton has a positive charge, denoted by e. A neutron is uncharged. A proton consists of two up quarks and a down quark, and a neutron consists of two down quarks and an up quark. Work out for yourself that an up quark has a charge of $+\frac{2}{3}e$ and a down quark has a charge of $-\frac{1}{3}e$. The quarks in a proton or a neutron are held together as a result of exchanging virtual particles called **gluons**. See Figure 11.2(i).

The strong nuclear force is a residual effect of gluon exchange in a proton or a neutron. This occurs as pion exchange between protons and neutrons when they get too close. When two protons or neutrons get too close, a gluon from a quark creates another quark-antiquark pair. The antiquark and the original quark form a pion which escapes to the other proton or neutron, leaving behind the new quark. The errant antiquark is annihilated by a quark in the other proton or neutron, leaving the escaped quark in a new home. See Figure 11.2(ii).

The weak nuclear force acts when an up quark suddenly becomes a down quark or vice versa. This change causes the release of an intermediate particle called a **W-boson** which disintegrates into a neutrino and a positron or an antineutrino and an electron. See Figure 11.2(iii). The W-bosons are the carriers of the weak nuclear force. These particles have been detected in particle collision experiments with the correctly predicted mass of 87 times the rest mass of the proton. How a quark which has a rest

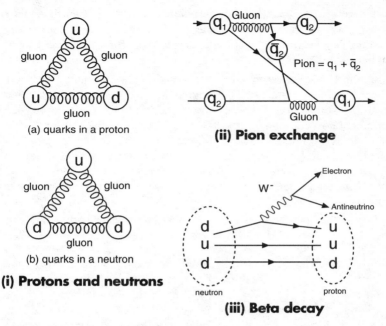

Figure 11.2 Quarks in action

mass less than that of the proton can emit a particle 87 times more massive than a proton is another story. Suffice to say that energy can be borrowed to create mass, and then 'paid back' again without detection, if the period of borrowing is sufficiently short. This is known as the **Uncertainty Principle** which ranks alongside $E = mc^2$ as one of the great theoretical discoveries of the twentieth century. We will return to the Uncertainty Principle in the final chapter of this book.

Matter and radiation

The implications of the quark model for cosmology were not immediately apparent when the model was first proposed. Why should events on the largest possible scale be affected by what happens or happened on the smallest scale? When atoms react chemically to release energy, the energy released is of the order of tens of millions of joules per kilogram. When nuclei fuse together to release energy, the energy released is about a million times greater, of the order of tens of millions of millions of joules per kilogram. The smaller the distances involved, the greater the energy released. The energy released in the Big Bang was so enormous that it created all the galaxies and made them speed away from each other to vast and increasing distances.

Most of the Universe except for the stars is now cold and empty, seemingly at an average temperature of a few degrees above absolute zero, according to the measurements of the microwave background. Temperature is a measure of how concentrated the thermal energy of an object is. In the Big Bang, the temperature when neutrons and protons formed from quarks would have been millions of millions of degrees, exceedingly high but nevertheless low enough to enable neutrons and protons to form. Before this era, the temperature would have been even greater, too great to enable neutrons and protons to form from quarks. The formation of quarks and antiquarks in high-energy accelerator experiments occurs when particles collide after being accelerated through billions of volts in massive high-energy accelerators. At such high energies, the colliding particles are massive enough to create matter and antimatter in the form of quarks and antiquarks. To achieve such energies thermally, the temperature of the particles would need to be raised to billions of billions of degrees. In the Big Bang, the formation of matter would not have occurred until the temperature had fallen below this level. Before that, the Universe would have been dominated by radiation at even higher temperatures.

Cosmologists began to pay more and more attention to the work of particle physicists after the acceptance of the Big Bang theory. From experiments involving particle collisions at higher and higher energies, it is now known that:

■ matter consists of **quarks** (which make up protons and neutrons) and lighter particles like electrons and neutrinos, known collectively as **leptons**. Antimatter consists of antiquarks and antileptons

■ quarks and antiquarks interact with each other by exchanging particles called **gluons** which can create pairs of quarks and antiquarks

■ leptons interact with each other and with quarks by exchanging **bosons** or, in the case of electrons, by exchanging photons

■ gluons, bosons and photons are force-carrying particles created and absorbed when quarks or leptons interact

■ quarks and leptons and their antiparticle counterparts can only be created or annihilated in particle-antiparticle pairs, unlike the force-carrying particles which don't obey such rules

■ a quark can change its identity by emitting a W-boson which then decays into an electron and an antineutrino or a positron and a neutrino.

Theorists reckon that the force-carriers become indistinguishable at sufficiently high energies, beyond the range of the current particle accelerators. The Large Hadron Collider is a massive billion-dollar project at CERN, the European Centre for Nuclear Research in Geneva. It is designed to probe the collisions of particles accelerated through billions of volts into each other. A single proton accelerated in this machine would have a mass thousands of times greater than its rest mass, enough to create quarks, antiquarks, bosons and gluons in abundance. A primary aim of these experiments is to find out why particles gain mass when they are accelerated. These experiments will give scientists a fleeting glimpse into the behaviour of matter at the enormously high temperatures, before matter formed from radiation in the Big Bang, when particles and antiparticles were produced and annihilated without any permanence. See Plate 7.

Volts, volts and more volts

Most people are familiar with the humble volt, the unit of voltage which is marked on electrical appliances and batteries. The energy of a single electron emerging from a 1.0 V battery is called an 'electron volt', abbreviated as eV. In a television tube, electrons are accelerated up to several thousand volts and then deflected by electromagnets to produce an image on the screen. The electrons hit the fluorescent coating inside the screen and make its atoms emit photons. In the Stanford Linear Accelerator, electrons are accelerated through voltages up to 50 000 million volts, each electron gaining 50 000 million electron volts (abbreviated as MeV) in the process. The rest mass of an electron, expressed in terms of energy, is 0.55 MeV and the rest mass of a proton is about 1000 MeV. An electron at 50 000 MeV would be capable of creating protons and antiprotons. The Large Hadron Collider is designed to accelerate particles to energies of the order of 7 000 000 MeV, enough to create bosons and other types of particle which only ever existed before in the Big Bang.

Particle energies and temperature

The temperature of a substance is a measure of the average kinetic energy of its particles. The particles of a gas at about a thousand degrees have energies of the order of about one tenth of an electron volt on average. The particles at the solar surface at 6000 degrees have energies of the order of about 0.5 eV on average. Nuclear fusion requires temperatures of the order of hundreds of millions of degrees, corresponding to particle energies of the order of MeVs and radiation wavelengths of the order of the diameter of the nucleus. At particle energies of the order of 50 000 MeV in the Stanford Linear Accelerator, the conditions correspond to matter at temperatures of the order of millions of millions of degrees, and to radiation wavelengths much smaller than the diameter of a proton. The Large Hadron Collider, designed to achieve particle energies a thousand or more times greater than the energies achieved in the Stanford accelerator, will be able to probe matter on a scale less than a ten-thousandth of the diameter of a proton. At these very small distances, the electromagnetic force and the weak nuclear force are equally strong and just as likely to act, either through the exchange of photons in the case of the electromagnetic force or through the exchange of bosons in the case of the weak nuclear force. In essence, the two forces are indistinguishable on this scale, unified as the **electroweak** force. The annihilation of an electron and a positron on this scale is just as likely to produce a boson as a photon, with both force-carriers just as likely to produce another particle-antiparticle pair subsequently. The Large Hadron Collider is designed to find out how bosons themselves interact and why they are so heavy.

The proton is the most stable form of matter composed of quarks. The possibility that the proton itself is unstable is a prediction of Grand Unification, a theory which unifies the strong nuclear force and the electroweak force. According to this theory, protons decay with a half-life of more than 10^{30} years, far far longer than the age of the Universe. At present, experimental evidence is inconclusive. Such a process would turn quarks into leptons and mesons. Maybe, in the very early stages of the Big Bang, the reverse process produced quarks and antiquarks. Perhaps, the Large Hadron Collider might provide some clues.

Here is a summary of how the properties of matter on the tiniest scale have affected events on the largest possible distance scale:

- The temperature of the Universe was of the order of billions of billions of billions of degrees when quarks and antiquarks formed permanently from high-energy radiation.
- As the Universe expanded, its temperature fell. Quarks combined in trios to form neutrons and protons when the temperature fell below a few million million degrees.
- Atoms formed when the Universe expanded enough for its temperature to fall below a few thousand degrees. Above this temperature, electrons break free from atoms and ions form.
- The Universe is now at an average temperature of just a few degrees above absolute zero. Galaxies which were formed from matter billions of years ago continue to rush away from each other.

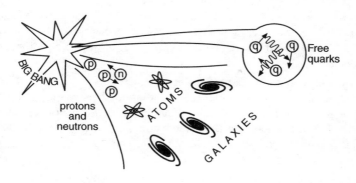

Figure 11.3   The formation of matter

Why is there more matter than antimatter?

Particles and antiparticles are formed in pairs from high-energy radiation. For every particle formed in a pair-production event, an antiparticle is also formed. These can only be detected in circumstances such as the debris produced from cosmic radiation or high-energy particle beam experiments or in the radioactive decay of proton-rich nuclei. Particles rather than antiparticles constitute the stuff of every substance we know. Perhaps billions of years ago, matter and antimatter swirled round in space in abundance, releasing radiation as particles and antiparticles annihilated each other. If there had been equal amounts of matter and antimatter formed in the Big Bang, it would be reasonable to expect neither matter nor antimatter to remain. Yet vast quantities of matter, not antimatter surround us.

Cosmologists reckon that there are about 10^{80} protons in the Universe, a colossal number which would take over a line on this page to express if written as 1 followed by 80 zeros. Every kilogram of matter contains about 10^{26} protons, and an average star of mass about 10^{30} kg therefore contains about 10^{56} protons. A galaxy containing about 10^{12} stars on average therefore contains about 10^{68} protons. Galaxies are spaced at distances apart of the order of megaparsecs. Because a volume of a cubic megaparsec is of the order of 10^{68} cubic metres, the Universe contains less than a proton per cubic metre on average. The distance to the furthest galaxies is of the order of 4000 megaparsecs, so there could be of the order of 10^{12} galaxies in the Universe. This gives about 10^{80} protons in the Universe.

How many photons are there in the Universe? At a background temperature of a few degrees, the radiation density of the Universe corresponds to an energy density of about 10^{-14} joules per cubic metre or about 400 million photons per cubic metre. For a Universe of the order of 10 000 billion light years in radius, the measured radiation density gives about 10^{89} photons in total.

The estimates above indicate there are about a thousand million photons for every proton. Protons represent energy locked up as matter after the Big Bang. Background microwave photons represent particle-antiparticle annihilations that happened billions of years ago. An imbalance of just one extra particle for every thousand million particles and antiparticles would be enough to create a proton-to-photon ratio of one to a thousand million. A millionth of a tenth of one per cent more particles than antiparticles is perhaps all that was needed to produce a Universe filled with matter rather than radiation or antimatter. Perhaps, in the primordial fireball conditions of the Big Bang, the laws of nature revealed a very slight preference for matter over antimatter or perhaps the rapidness of the expansion caught the Universe in an uncertain state when there was more matter than antimatter.

ACTIVITY

A game of chance

1 Shuffle a pack of playing cards containing equal numbers of red and black cards and then spread the cards out face down at random. The red cards are to represent quarks and the black cards to represent antiquarks.

2 Pick up three face-down cards.

- ■ If you get three red cards, put them on one side together as a 'matter particle'.
- ■ If you get three black cards, put them together on the other side as an 'antimatter particle'.
- ■ If you haven't got either of the above, put a red and a black card from the trio together in the middle as a 'photon' formed by a quark and an antiquark annihilating each other. Put the spare card face down among the other cards and 'stir' them all round a bit.

3 Repeat the procedure in 2 until all the cards are in sets, either in the matter group or the antimatter group or in the photon group. There may be one or two cards left over.

4 Count up how many matter particles you created, how many antimatter particles you created and how many photons were created.

Statistically, there are 8 possibilities for each trio, namely RRR, RRB, RBR, BRR, RBB, BRB, BBR and BBB. For every matter particle (i.e. RRR), there should be one antimatter particle (BBB) and six photons (from RRB, RBR, BRR, RBB, BRB, BBR). However, the law of chance is at work and you could end up with more matter than antimatter or vice versa in different parts of the expanding Universe. Had this happened, regions of matter and regions of antimatter would have been produced which would have annihilated each other where they were in contact with each other, releasing gamma photons in the process. No evidence has been found to date that such processes occurred and there seems to be an excess of matter over antimatter in the Universe as a whole. Another possible explanation is that mesons comprising a strange quark and an up or down antiquark formed. These mesons, known as K-mesons, then decayed to produce quarks, antiquarks, leptons and antileptons with a slight bias towards matter particles. Evidence for a natural bias in the decay of K-mesons into pions, leptons and antileptons was discovered in 1964 by two American physicists, Jim Fitch and Val Cronin. The weak nuclear force is at work in this type of decay, and Fitch and Cronin discovered some decays unexpectedly broke the rules of symmetry by producing no leptons or antileptons, and producing two pions per decay only. Fitch and Cronin shared the 1980 Nobel Prize for physics for their discovery. If the weak nuclear force and the strong nuclear force are shown to be unified at very high energies, the unified force might also display a bias towards quarks rather than antiquarks. The idea offers a way out of the conundrum of why there is more matter than antimatter in the Universe. Tests are now underway to see if there is a natural bias towards matter in the decay of B-mesons which are formed from heavy strange quarks called 'bottom' quarks and antiquarks. These are created in high-energy collisions between electrons and positrons.

Does gravity have a place in a unified theory of force? At exceedingly high energies, a combined electroweak and strong nuclear force might merge with the force of gravity. The force of gravity could be carried by the so-called **graviton**, a packet of gravitational waves extending over enormous distances in space, travelling through space at the speed of light. In the Big Bang, when the distance scale was exceedingly small and radiation energies were exceedingly large, perhaps gravitons were indistinguishable from photons, bosons and gluons. A grand unified theory of forces, GUTS for short, if proven, would be worth a Nobel prize or two!

ACTIVITY
Another game of chance
Repeat the game on the previous page with an extra red card. You should find you end up with more matter than antimatter every time!

Summary

Forces

- ■ **Electromagnetic forces** act between charged objects due to the exchange of virtual photons, repel if the charges are like, attract if unlike, become weaker with distance, and have infinite range
- ■ **Gravitational forces** act between objects due to their mass, are attractive only, become weaker with distance and have infinite range
- ■ **Strong nuclear forces** act between neutrons and protons due to the exchange of pions, have a range of about 10^{-15} metres, are attractive except at very short range and act equally between neutrons and protons
- ■ **Weak nuclear forces** are responsible for beta decay processes in which a boson is created and decays into a lepton and an antilepton over a range of no more than about 10^{-18} m

Matter and antimatter

Matter exists as either quarks or leptons. A proton consists of two up quarks and a down quark. A neutron consists of two down quarks and an up quark. The lepton family consists of electrons and neutrinos. Antimatter consists of antiquarks and antileptons.

In the Big Bang

As the Universe expanded and cooled,

- ■ radiation turned into matter and antimatter which turned back into radiation at random
- ■ quarks and leptons (electrons and neutrinos) formed from radiation
- ■ protons and neutrons formed from quarks
- ■ nuclei formed from protons and neutrons
- ■ atoms formed from nuclei and electrons

12 THE EARLY UNIVERSE

The most distant quasars are known to be billions of light years away. Light from such an object reaching us now set off on its long journey through space billions of years ago. We see the quasar as it was when the light now reaching us from it was emitted. The more powerful a telescope is, the fainter the quasars that can be observed with it, enabling us to observe objects further and further into the distant past. Cosmologists can therefore study the Universe as it was billions of years ago by observing quasars billions of light years away. In this chapter, we shall look at what has been discovered about the early Universe as a result of using telescopes to observe quasars and other objects billions of light years away. In addition to using optical telescopes, such studies have been carried out using ground-based radio telescopes and ultraviolet and infra-red cameras fitted to telescopes carried by satellites. Recent observations have led cosmologists to rethink how galaxies formed and how they develop. Evidence for violent events before galaxies formed has also been gathered in recent years. Whilst cosmologists are attempting to look further and further back into the past, theoretical models are being developed to try to understand how the Universe evolved from the primeval fireball of the Big Bang to its present state. We will look at why cosmologists reckon that the rate of expansion of the Universe dramatically increased for a short time at a very early stage, and we will look at how matter in its present state formed afterwards.

Images from the past

Quasars have presented astronomers with many unsolved problems ever since the first quasar was discovered in 1963 by Maarten Schmidt at the California Institute of Technology. A quasar is a point-like source that produces radiation with a large redshift (i.e. a large increase in the wavelength of its light due to its receding motion). Variations in the light intensity from a quasar over little more than a few days indicate that quasars are much smaller than galaxies, which are thousands of light years in diameter. The size of a quasar cannot be much more than the distance travelled by light in a few days, if intensity variations over a few days are detected from it. This sort of effect is at work if you have ever listened to a very large crowd at a football stadium attempting to sing the same song. Because the time taken for sound to travel across the crowd is of the same order as the variations of intensity associated with the song, the singing becomes indistinct as the variations even out into a tuneless drone. The larger the crowd, the more indistinct the singing. An object of the order of thousands of light years in size would be far too large to produce light variations of the order of days.

The rate of release of energy from a quasar is enormous, far greater than from the most powerful stars. On page 133, we met the possibility that a massive black hole at the centre of a galaxy could be the cause of the enormous power output of a quasar, drawing in surrounding stars from the galaxy and releasing powerful radiation as the stars are sucked into the black hole. The fact that quasars produce large redshifts means that they are moving away from us at very high speeds. According to Hubble's Law, they are therefore very far away, in fact billions of light years away from us.

■ The red shift of a quasar or any other receding object is defined as the percentage change of the wavelength of light from it divided by 100. For example, if the wavelength is increased by 50%, the red shift is 0.5. This means that the observed wavelength is 50% longer than the emitted wavelength. The table below shows the red shift for various percentage changes of wavelength. The largest red shift of a quasar is about 4 which means its wavelength is stretched by 400%. In this case, light that would have a wavelength of 0.5 micrometres is stretched 2.0 micrometres to a wavelength of 2.5 micrometres in the infra-red region of the spectrum by the receding motion of the quasar.

■ If the red shift is much less than 1, the speed of the receding object is equal to the red shift × the speed of light. For example, an object with a red shift of 0.1 is moving away at a speed of 0.1c, where c is the speed of light in space. Hubble's Law can then be applied to work out the distance to the object. For an object which has a red shift of 0.1, its distance is therefore equal to $0.1 \, c/H$, where H is the Hubble constant. Since $c = 300\,000$ km s^{-1}, then this distance is equal to 430 Mpc ($= 0.1 \times 300\,000/70$) assuming $H = 70$ km s^{-1} Mpc^{-1}. As 1 parsec equals 3.26 light years, a distance of 430 Mpc is approximately equal to 1400 million light years or 1.4 billion light years.

■ If the red shift is larger, the distance cannot be worked out using the previous method. This is because time dilation is significant, and space-time cannot be assumed flat over such large distances. For much smaller distances, space-time is effectively flat just as a small patch on a football is almost flat. However, just as a large patch on a football is curved, perhaps space-time is curved on a large enough scale. Calculation of the distance to a quasar from its red shift for larger redshifts therefore requires assumptions to be made about the link between space and time as the Universe expanded.

Cosmic time

Imagine you had a time machine that enabled you to move about in space-time so you could be anywhere in the Universe at any time. If you travelled back to the early Universe, you might perhaps see quasars much closer than we see them or, even earlier, you might witness how clusters of galaxies formed and whether or not galaxies formed inside clusters or clusters formed from galaxies. Your time machine could be fitted with a cosmic clock to tell you which 'era' the machine has taken you to. In fact, such a clock would not really be needed as you would only have to observe the local region to tell which era you were in. For example, if you observed lots of quasars nearby, you would know that you had travelled back in time billions of years. The concept of 'cosmic time' is therefore based on the idea that any two observers in different parts of the Universe, who see the same overall picture of the Universe, are observing at the same cosmic time.

A forbidden journey

Time travel into the past is thought to be forbidden because it would violate the basic principle of cause and effect. Imagine one of your grandparents had been very wealthy and had disinherited one of his children, your parent, before you were born. You return to the past to a time before your parent was born to remonstrate with your mean grandparent. In the ensuing row, your grandparent flies into a rage and dies of a heart attack. Your parent would not therefore have existed and neither should you. Time travel into the past would allow an effect to happen before its cause in violation of the basic principle of cause and effect.

Scaling the Universe

Hubble's Law tells us that the distant galaxies are moving away from each other. This is thought to be due to the expansion of the Universe which began with the Big Bang. This expansion has caused the distances between the galaxies to increase without altering the size of the galaxies. In other words, the expansion of the Universe is making the galaxies move apart without making them bigger. Light travelling to us from a distant galaxy is stretched in wavelength by the expansion of the Universe as it travels through space, causing its red shift. The wavelength of such light is longer when it reaches us than when it was emitted.

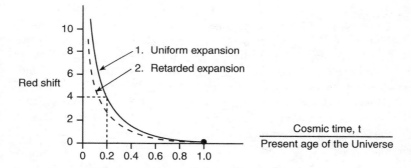

Figure 12.1   Red shift and cosmic time

Figure 12.1 shows how the red shift increases with time back to the Big Bang.

1 The full-line curve shows how the red shift varies with the time when the light was emitted, assuming the wavelength of light increases steadily with cosmic time. This corresponds to a Universe which has expanded uniformly since shortly after the Big Bang. For example, a red shift of 1 corresponds to an emitted wavelength half the wavelength we observe (i.e. a 100% increase of wavelength) and therefore a Universe half its present size and age. In comparison, a red shift of 4 corresponds to an emitted wavelength one-fifth the wavelength we observe (i.e. a 400% increase of wavelength) and therefore a Universe one-fifth its present size and age.

Red shift and cosmic time

In this model, the time measured from the Big Bang, $t = \dfrac{1}{(1+z)} \times T$,

the present age of the Universe, where z = the red shift of the quasar that emitted the light. Assuming the wavelength increases in proportion to cosmic time, the redshift z is therefore equal to the ratio $\dfrac{(T-t)}{t}$

where t is the cosmic time when the light was emitted and T is the present age of the Universe.

---

2 The dotted line shows how the red shift varies with cosmic time for a Universe in which the expansion is slowing down due to gravity. In effect, the ⅔ factor due to gravity from page 164 comes into the situation, so the time since the light was emitted is increased by a factor of ½. Clearly, the deduction of distance from redshift for large redshifts depends on assumptions about the rate at which the Universe has expanded. More advanced theoretical models based on the General Theory of Relativity give the same broad spread of possible distances, depending on the assumptions made.

The effect of gravity on the expansion of the Universe is not known for certain because the amount of matter in the Universe is not known. Considerable research is under way in the search for sources of dark matter and other forms of mass. We will return to the subject of missing mass in Chapter 13 when we consider the future of the Universe.

The age of quasars

Over the past two decades, many quasars have been discovered and measured for redshift and therefore for distance. The general pattern of the measurements shows that the number of quasars is greatest at a red shift of about 2 which corresponds to a distance of between 8 and 10 billion light years. In addition, very few quasars are found at a red shift of more than about 4 which might mean that quasars did not come into existence until a few billion years after the Big Bang. Figure 12.2 shows how the number of quasars at different distances varies with redshift.

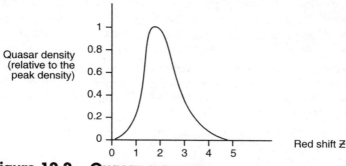

Figure 12.2   Quasar surveys

The fact that the number of quasars peaks at redshift 2 is evidence of a 'quasar age', about 8 billion years ago. Perhaps this era was a time of galactic chaos when galaxies were much closer and were colliding with each other, maybe smaller galaxies being swallowed up by large galaxies, feeding black holes lurking in the galactic centres of large galaxies. Quasars cannot have formed much earlier as they would be observed at much higher redshifts. The possibility that quasars are continuing to form is effectively ruled out as the most powerful radio galaxies nearby are known to be little more than a few million years old, much younger than the billion years or so to count as a quasar still warming up. In addition, although quasars cannot be resolved into individual stars, distant radio galaxies are found to contain much younger stars than nearby elliptical galaxies, indicating that these radio galaxies were in the process of formation when they emitted the light which we are now observing from them.

Gamma ray bursts

If your eyes could detect gamma rays, you would be surprised to observe flares in the sky occasionally, each perhaps lasting as long as a minute or so. In the 1960s, the US Defense Department launched a series of satellites to spot any secret nuclear weapons tests in space by the USSR. Instead, the satellites unexpectedly discovered gamma ray bursts from random directions in space. The cause of these gamma flashes remained a source of speculation among astronomers until 1991 when a satellite carrying a gamma ray observatory was launched into space from the space shuttle Atlantis. The observatory detected bursts at a rate of about one per day from random directions. This discovery convinced astronomers that gamma ray bursts originate outside the Milky Way galaxy, otherwise more bursts would have been detected from the direction of the Milky Way than from other parts of the sky. The randomness of the locations of the gamma ray bursts means that the phenomenon must be associated with the Universe at large, not with any particular part of the Universe. The possibility that the bursts could be due to neutron stars in the halo surrounding the Milky Way galaxy was ruled out since no such distribution was detected surrounding our sister galaxy, Andromeda.

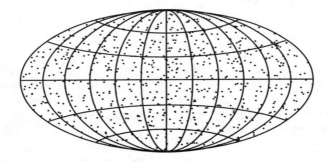

Figure 12.3   Sky map of gamma ray bursts

The problem of pinpointing a gamma ray burst to identify its precise source was beyond the specification of the gamma ray observatory. A new satellite, Beppo-SAX, was launched in 1996 which included a wide-angle X-ray telescope on board designed by a team of Italian and Dutch astronomers. In February 1997, the gamma detector on the satellite was triggered by a gamma ray burst, designated GRB 970 228, that lasted over a minute from the direction of Orion. The team was able to pinpoint the source using the X-ray telescope within eight hours when the source was still bright enough to detect. The result was confirmed within twelve hours by astronomers using the 4.2 m William Herschel Telescope at La Palma in the Canary Isles. Later observations using the Hubble Space Telescope revealed a fading point object surrounded by an unchanging fuzzy glow, thought to be the home galaxy of the point object. Beppo-SAX located another gamma burst, GRB 970 508, in May 1997, which proved to be even more revealing than its earlier discovery. This time, the source was rapidly located by astronomers using the 0.9 m telescope at Kitt Peak, Arizona who discovered it became brighter the next night, reaching peak brightness four days after it was first detected. This four-day period gave astronomers using the 10 m Keck telescope on Hawaii sufficient time to record the optical spectrum of the source. This was found to include absorption lines due to magnesium and iron at a red shift of 0.84. At last the question was settled about whether or not gamma ray bursts are near or far as a red shift of 0.8 clearly corresponds to a distance of the order of billions of light years. More significantly, the faintest gamma ray bursts recorded could be ten billion or more light years away at redshifts of the order of 6 or more, beyond the range of the most distant quasars.

The cause of gamma ray bursts is not yet known. Perhaps gamma ray bursts are the death throes of massive stars formed a billion years or so after the Big Bang, maybe before galaxies formed in the 'dark age' of the Universe between the end of the radiation era when atoms and ions formed and when galaxies began to form. The energy released by a gamma ray burst is reckoned to be of the same order as that which a supernova releases. However, a supernova remains bright for months or even years whereas a gamma ray burst fades within minutes. This could be a time dilation effect or it could be on account of energy being suddenly released as jets of matter and radiation from a massive supernova. Whatever the cause of a gamma ray burst proves to be, we know such bursts happened before the age of quasars when the Universe was no more than about a billion years old.

Inflation at work

The horizon problem

As outlined on page 170, crucial evidence for the Big Bang was gathered by Penzias and Wilson in 1965, when they detected microwave radiation from all directions in space. This radiation was found to be remarkably uniform, from all directions at a temperature of 2.7 degrees above absolute zero. In 1989, the Cosmic Background Explorer (COBE) satellite mapped this radiation out and discovered very slight unevenness corresponding to temperature variations of the order of millionths of a degree.

The microwave background radiation is radiation released billions of years ago when neutral atoms formed from ions and electrons. This radiation has been stretched out by the expansion of the Universe and was much more energetic when it was released. Ions and electrons formed neutral atoms as the Universe expanded when the energy of the background radiation lessened enough to allow atoms to form. This would not have happened until the temperature of the Universe had fallen below about 4000 degrees. At this stage, matter and radiation became 'decoupled' and the Universe became transparent. Radiation released then has been travelling through space ever since, being gradually stretched to longer and longer wavelengths until it is now in the microwave region. The Universe was 1000 times hotter when the radiation was released and therefore about 1000 times smaller.

The discovery that the temperature associated with the microwave background radiation is the same in all directions presented a problem to the Big Bang theory. A hot object and an identical cold object in an insulated empty box will eventually be at the same temperature because the hot object initially radiates more energy per second than it receives from the cold object whereas the cold object receives more than it radiates. So the hot object loses thermal energy and the cold object absorbs thermal energy until they are both radiating and absorbing thermal energy at the same rate. However, if one object is ten billion light years in one direction and the other object is ten billion light years in the opposite direction, they cannot possibly have reached thermal equilibrium ten billion years ago since the radiation from each is only just reaching us. Yet they are at the same temperature.

Figure 12.4(a)   The horizon problem

Figure 12.4(a) shows a different perspective on this problem. Since nothing can travel faster than light, we can see objects no further than about 10 billion light years away, corresponding to when the Universe

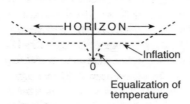

Figure 12.4(b) Inflation at work

became transparent. Our horizon is therefore defined in Figure 12.4 by a zone stretching back in time to a radius of about 10 billion light years. Objects in opposite directions that released radiation about a billion years after the Big Bang would not have had time to exchange thermal radiation as their horizons would not have overlapped. The problem became known as **the horizon problem**.

Another problem with the Big Bang model, in its original form, is that it could not explain why the average density of the early Universe was so close to the **critical density**, which is the least density that would allow it to expand to infinity without collapsing back under its own gravity. A lesser density at this stage would have produced a present-day Universe much much less dense than we observe. A greater density would have led to a collapse by now. This problem is known as the **flatness problem** as the critical density corresponds to a universe with zero curvature of space-time whereas a larger density corresponds to a universe in which space-time curves back on itself. We will return to this question of the critical density and flatness later in this chapter.

Both the horizon problem and the flatness problem were solved by introducing a new idea into the Big Bang model. This idea, known as **inflation**, was introduced by an American physicist, Alan Guth, in 1981. According to Guth, the Universe underwent extremely rapid expansion by an enormous factor in a very short period at a very early stage in its expansion, about 10^{-35} seconds after the Big Bang. Guth worked out the expansion factor was of the order of 10^{50} in a tiny fraction of a second during which a permanent excess of particles over antiparticles was created. The expansion continued much more slowly after its inflationary phase, taking 100 000 years to expand to about a thousandth of its present size when it was sufficiently cool to become transparent. Ten thousand million years later, the Universe has expanded by a factor of a thousand or so and here we are! Guth worked out that a brief phase of rapid inflation allowed enough time for the temperature of the Universe to become uniform before the rapid inflation occurred. This theory solved the horizon problem. In other words, the uniformity of the Universe was established in its relatively slow expansion before inflation. Once established, this uniformity remained unaffected by inflation.

Inflationary economics

Inflation is a runaway process in which the rate of increase increases and increases. Work out for yourself the cost of a loaf of bread if its price doubled every month for a year from 50p at first. The increase of 50p per month at first would rise to £1024 within 12 months.

Suppose in a thousand years' time, there is a single world currency and the price of any article is the same in every part of the world. Over the preceding ten centuries, inflation increased prices by a factor of a million yet prices in the 29th century are uniform everywhere, increasing at a much slower rate than the average for the previous nine centuries. This scenario involves equalization of prices and inflation of prices. Which one of the following four sequences is most likely to have occurred over these ten centuries?

1 Early equalization followed by steady inflation
2 Early equalization followed by rapid inflation followed by slow inflation
3 Steady inflation followed by late equalization
4 Rapid inflation followed by late equalization then slow inflation.

Scenarios 1 and 3 would be ruled out because inflation was not steady. Scenario 4 would lead to a wide divergence of prices before equalization with insufficient time for equalization. The second scenario fits the situation, just as the inflationary model of the Big Bang explains equalization of temperature.

Wrinkles and ripples

The density of the Universe today is reckoned to be about 10^{-27} kilograms per cubic metre, corresponding to no more than a proton per cubic metre on average. This is a little less than although of the same order of magnitude as the critical density which would be just sufficient to allow the Universe to expand to a standstill at infinity without collapsing back. The ratio of the average density of the Universe to its critical density is called the **density parameter**, designated by the greek letter Ω (pronounced 'omega'). Since the average density at present is close to the value of the critical density, the present value of Ω is therefore about 1.

The Universe has expanded enormously since the end of the inflationary era from the size of a golf ball to its present size of the order of ten billion light years, a factor of about 10^{28} times larger. With or without inflation, the enormous expansion of the Universe would have produced a density parameter very different to its present value if its value in the early Universe had differed from unity by more than about 1 part in 10^{28}. Had the density parameter differed from unity by a tiny fraction, the difference would have been magnified enormously to give a present-day density parameter well outside its measured range.

Why was the density parameter so very close to unity in the early Universe? The original non-inflationary model of the Big Bang provides no explanation. Guth worked out that a burst of inflation 10^{-34} s after the Big Bang lasting a hundred times longer forced the density parameter to unity to within one part in 10^{50}. Guth's inflationary model of the Big Bang thus provides an explanation of why the density parameter of the Universe is still close to 1.

Inflation also explains why the temperature of the Universe is so even, varying by less than 30 millionths of a degree against a background temperature of 2.7 degrees above absolute zero. This **smoothness** is because the pre-inflationary Universe was small enough to become uniform as each part was able to exchange radiation with every other part. As explained on page 128, the distribution of clusters of galaxies on the scale of the Universe is even to within one part in 100 000 which is about the same as the variation of temperature of about 30 millionths of a degree to 2.7 degrees. Inflation and its aftermath has expanded the scale of the Universe enormously but seems not to have altered the smoothness of the Universe. The ripples in the microwave background, detected by COBE, represent wrinkles due to small-scale fluctuations in the Universe at its inflationary stage. See Plate 10.

The mechanics of inflation

The Big Bang theory is supported by experimental evidence as outlined in Chapter 10 and is underpinned by Einstein's General Theory of Relativity. Inflation was introduced to account for the fact that the microwave background radiation is *isotropic* which means it is uniform in all directions. We cannot rule out the possibility of alternative explanations for this isotropy, such as a much much faster speed for light in the early Universe, but at the present time, most cosmologists accept the inflation model of the Big Bang. We saw in Chapter 10 that Einstein came very close to inflation and the Big Bang theory when he found that a static Universe could only be stable by introducing a repulsive force to counteract the attractive force of gravity. This repulsive force is defined by means of a cosmological constant λ (pronounced 'lambda') which needs to be positive for a repulsive force, zero for zero force or negative for an attractive force. Einstein thought the Universe is static and therefore needed to prevent gravity causing its collapse so he made λ positive to counteract gravity. Other mathematicians later showed that Einstein's General Theory of Relativity produces an 'expanding Universe' solution without the need for this extra force. Although Einstein regarded his introduction of the cosmological constant as a blunder, his idea has been revived as the mechanism for inflation. If this cosmological repulsive force was dominant in the very early Universe, it would have caused an exponential growth in the scale of the Universe sufficient to explain the inflationary era. Regardless of whether or not a cosmological repulsive force was present in the early Universe, there is little direct observational evidence for its existence now. If the most distant objects in the observable Universe are still accelerating, it would be reasonable to expect their redshifts to increase more with distance than for nearer objects. Measurements of the red shifts of distant supernovae suggesting acceleration are at present inconclusive.

Inflation is thought to have forced the density parameter Ω to become unity. Alexander Friedmann, the originator of the 'expanding Universe' solution to Einstein's equation, showed that the density parameter differs from unity by a factor which depends on the curvature of space. The flatter the curvature of space, the smaller this factor is. During the inflationary era, the density parameter was forced to unity as the Universe expanded dramatically and became much flatter. We shall see in Chapter 13 that the curvature of space determines whether or not the Universe will continue to expand for ever or collapse to a big crunch!

Where did the energy needed to drive the inflationary era come from? This key question is unanswered at the present time. During the inflationary era, energy locked up in space was suddenly released as space unfurled itself and became flatter. As we saw in Chapter 9, gravity distorts space, so the flattening out of space during the inflationary era perhaps released gravitational energy, flinging the Universe into a brief bout of accelerated expansion. Particle physicists hope to learn more about the nature of space, energy and mass from collision experiments using high-energy particle beams.

A flatness test

Inflate a balloon partially and draw a grid of lines on a small part of its surface with a suitable pen. For a spherical balloon, the curvature of each line is defined as $1/R^2$, where R is the radius of the circle the line would form if it was continued all the way round the balloon. Now inflate the balloon fully, and you will see that the lines are less curved and the balloon surface is flatter. Zero curvature corresponds to a completely flat surface.

Imagine the balloon skin briefly weakened as it was expanding. This would enable it to expand very rapidly until it regained its strength and the accelerating expansion stopped.

The history of the early Universe

The expanding Universe became transparent when its temperature had fallen below about 4000 degrees, the temperature above which electrons in atoms can gain enough energy to break free when atoms collide. The surface of a red star is at about 3000 degrees, heated by a steady stream of high-energy radiation from the core of the star. At the surface, atoms are broken apart into electrons and ions as they collide violently with each other or they absorb high-energy radiation directly. Light is emitted when an ion and an electron recombine or when electrons fall deeper into the atom. The expanding Universe was a tiny fraction of its present size when most of its electrons and ions finally recombined, unable to break away from each other as the radiation was stretched and weakened too much to break the atoms up. The collisions between atoms became less violent, and the atoms could no longer break each other up on impact. At this stage, electromagnetic radiation stopped interacting with matter, as it is largely unaffected by neutral atoms, and so it set off on its long journey through space.

The Universe became transparent about a hundred thousand years after the Big Bang, when it was a hundred thousand times younger than it is now. Its very early phase of rapid inflation had caused it to expand to one thousandth of its present size, large enough to stretch its electromagnetic waves and prevent them from interacting with matter. What is thought to have happened before the Universe became transparent?

1 Quantum effects dominated the Universe for a very short time after the Big Bang

Particles and antiparticles appeared and disappeared in accordance with the rules of quantum physics, like fish jumping out of the sea. As the Universe expanded, its energy density decreased and quantum effects ceased to dominate the Universe when it was no more than about 10^{-43} s old. This estimate of time is known as the **Planck time** as it corresponds to the transition from quantum gravity to gravity associated with stable masses. The horizon of the Universe at this time would have been of the order of 10^{-35} m, corresponding approximately to its age of 10^{-43} s multiplied by the speed of light (3×10^8 m s^{-1}).

Estimating the Planck time

One of the most important ideas developed from Planck's quantum theory is that particles (and antiparticles) have a wave-like nature as well as an obvious particle-like nature. For example, a beam of electrons aimed at a very thin crystal is diffracted by the crystal in much the same way as a beam of X-rays. The idea of wave-like behaviour of particles was put forward in 1926 as a hypothesis by Louis de Broglie, a French physicist. He proposed that the wavelength of matter waves, now referred to as the de Broglie wavelength, is equal to h/p, where p is the particle momentum (defined as its mass $\times$ its speed) and h is the Planck constant, introduced on page 51. Because the value of h is so small at 6.6×10^{-34} J s, wave-like effects can only be produced with atoms and sub-atomic particles.

The Planck time can be estimated by equating the gravitational potential energy of two identical particles at separation r $(= Gm^2/r)$ to their relativistic energy $2mc^2$:

$$\frac{Gm^2}{r} = 2mc^2$$

At the transition, the de Broglie wavelength of either particle is approximately equal to half their separation $\frac{1}{2}r$, so the momentum of either particle is $2h/r$. Since the particle speeds would have been less than the speed of light c, the mass of either particle would be at least $2h/cr$

$\therefore$ for $m = \dfrac{2h}{cr}$, combining the two equations therefore gives $r^2 = \dfrac{Gh}{c^3}$

The length $r = [Gh/c^3]^{1/2}$ is determined by the three fundamental constants G, h and c which makes it a distance scale indicator below which quantum effects are dominant.

The Planck time $t = r/c$, corresponding to the time taken for the two particles to reach separation r from zero at speed c, is therefore a natural time scale for events dominated by quantum effects.

Hence $t^2 = Gh/c^5$

Since $G = 6.7 \times 10^{-11}$ N m^2 kg^{-2}, $c = 3.0 \times 10^8$ m s^{-1} and $h = 6.6 \times 10^{-34}$ J s, then $t^2 = 6.7 \times 10^{-11} \times 6.6 \times 10^{-34}/(3.0 \times 10^8)^5$

$\therefore t = 10^{-43}$ s

During this brief period after the Big Bang, quantum fluctuations dominated the Universe. After this period, quantum fluctuations ceased to dominate the Universe.

2 The inflationary era

The expansion of the Universe gathered pace after the quantum era, gradually at first then at an accelerating rate until suddenly it expanded by an enormous factor, rapidly running out of the necessary energy to maintain this sudden burst against the braking effect of gravity. This sudden expansion is thought to have been due to a repulsion effect driving matter apart. The effect is predicted by the General Theory of Relativity and leads to the result that the acceleration is in proportion to the radius of the Universe on a time scale of the order of 10^{-34} s. The result is that the Universe expanded by a factor of 10^{50} in less than 10^{-32} s.

The energy density of the Universe at this stage is reckoned to have been of the order of 10^{92} joules per cubic metre, corresponding to a mass density of about 10^{75} kilograms per cubic metre. The entire mass of the Universe at the end of this stage was probably packed into a sphere no bigger than a golf ball. Before this period of expansion, the temperature of the Universe would have equalized throughout as it cooled. The pressure inside the expanding Universe would have been maintained as more and more energy was released, expanding the Universe at an increasingly faster and faster rate. The Universe would have expanded one hundred-fold or so every 10^{-34} seconds until it had inflated in size by a factor of 10^{50}. Since an increase by a factor of 10^{50} is the same as twenty-five hundred-fold increases ($= 100 \times 100 \times 100 \ldots$ twenty-five times), this enormous bout of inflation lasted no longer than about 10^{-32} seconds, brought to an end by the supercooling effect of this rapid expansion causing particles and antiparticles suddenly to form.

The slight preponderance of matter over antimatter is thought to have occurred at this stage in a process known as **baryogenesis** in which massive particles and antiparticles formed at enormous energies corresponding to temperatures of the order of 10^{26} degrees. These particles and antiparticles decayed into quarks and leptons slightly more than into antiquarks and antileptons. The overall effect is an excess of quarks over antiquarks in the proportion of 1 extra quark to every 100 million quark and antiquarks produced. Yet this tiny imbalance is enough to account for all the matter in the Universe. The Grand Unified Theory (GUT) of forces shows this imbalance arises at temperatures below about 10^{26} degrees when the strong nuclear force and the weak nuclear force separate out.

Cooking up the quarks

Before baryogenesis, particles and antiparticles were confined to a horizon of about 10^{-28} m in size. The momentum of such a particle would therefore be of the order of about 10^{-5} kg m s^{-1}, corresponding to a de Broglie wavelength of about 10^{-28} m. The energy of a particle at this momentum is of the order of 1000 J (= momentum $\times$ speed of light) which is about the same as the energy you expend when you get out of bed. This energy is about 10^{24} times greater than that of a molecule of air at room temperature (which is about 300 degrees above absolute zero). Thus baryogenesis occurred when the temperature of the Universe fell below 10^{26} degrees (= 300×10^{24} approximately).

3 The free quark era

Free quarks no longer exist, even though every proton and neutron is composed of three quarks. After the inflationary era, quarks and antiquarks annihilated each other producing gamma photons until no more antiquarks remained. Until the Universe cooled sufficiently, the remaining quarks still had too much energy to combine with each other. Particle beam experiments involving head-on proton–proton collisions show that quarks can't be prised out of protons (or neutrons) unless the protons are moving towards each other at speeds in excess of 0.5c, where c is the speed of light. At such speeds, the kinetic energy of a proton is significantly greater than its rest-mass energy and enough kinetic energy is available to create new quark-antiquark pairs in a proton–proton collision. The temperature of a gas of protons would need to be in excess of 3 million million degrees for the protons in such a gas to be so energetic as to break each other apart on colliding. From this analysis, it follows that protons and neutrons could not have formed from free quarks until the temperature of the expanding Universe fell below about 3 million million degrees.

How long did the free quark era last?

The inflationary era ended as gravity started to hold back the runaway expansion of the Universe. The density of the Universe started to thin out as it expanded, due partly to its increasing volume and also to stretching its electromagnetic waves. It can be shown that the speed of expansion decreased in inverse proportion to the radius and that the time taken therefore was proportional to the square of the radius.

Since the temperature dropped as the radius increased, the time taken was therefore inversely proportional to the temperature squared. This relationship can be expressed in the form

$$\text{time} \times \text{temperature}^2 = \text{constant}$$

∴ time at the end of the free quark era $\times (3 \times 10^{12})^2$ = time at the end of inflation $\times (10^{26})^2$

Hence the time at the end of the free quark era = 10^{-5} s, since the time at the end of inflation = 10^{-32} s.

Since the temperature and hence the energy density of the Universe fell during the free quark era by a factor of about 10^{14} degrees (from about 10^{26} degrees to about 10^{12} degrees), the wavelength of its photons was stretched by the same factor. Thus the radius of the Universe must have increased by a factor of about 10^{14} in this era. During the free quark era, which lasted no more than about 10 microseconds, the Universe expanded from about the size of a golf ball to about the size of the Solar System and cooled to about 3 million million degrees.

4 Beyond the free quark era

Protons and neutrons formed from quarks when the Universe cooled below about 3 million million degrees. High-energy leptons and antileptons in the form of electrons, positrons, neutrinos and antineutrinos interacted with the protons and neutrons, causing protons to change into neutrons and neutrons to change into protons. When the temperature fell below 10 000 million degrees, the neutrinos and antineutrinos decoupled from these interactions due to their reduced energies.

5 Helium formed from hydrogen when the Universe was a few minutes old

As the Universe continued to expand, its temperature continued to fall until its protons and neutrons were moving slowly enough to combine into helium nuclei and other light nuclei. This process is known as **nucleosynthesis**. Before this point in time, the protons and neutrons possessed too much kinetic energy to stick together when they collided. The lightest nucleus of all the elements is the hydrogen nucleus which is a single proton. The next lightest nucleus is the helium nucleus which consists of two protons and two neutrons. A temperature of about 1000 million degrees or more is needed to make protons and neutrons fuse together, so the temperature of the Universe must have cooled to below 1000 million degrees when its helium and other light elements formed. The proportion of helium to hydrogen was set at this time to about 1:3. The successful explanation of this helium-to-hydrogen ratio is one of the main reasons why the Big Bang theory was accepted. See page 170.

The temperature of the Universe at nucleosynthesis was about 3000 times lower than at the end of the free quark era. The Universe must therefore have stretched by a factor of about 3000 to reduce its photon energies by a factor of 3000. The size of the Universe was therefore about 3000 times the size of the Solar System, of the order of 1 to 10 light years across when nucleosynthesis commenced. Using the same link between time and radius as before, the Universe must therefore have been about ten million times (= 3000 × 3000 approximately) older than at the end of the free quark era. Nucleosynthesis therefore started about a minute or so after the Big Bang.

6 The Universe suddenly became transparent about 100 000 years after the Big Bang

For many thousands of years after nucleosynthesis, light could not travel across the Universe because it was absorbed by matter in the form of free charged particles. For the same reason, light is unable to pass through a metal plate: the metal contains free electrons which absorb light at any frequency. At this stage, after the formation of small nuclei, the Universe was filled with a gas of charged particles in the form of nuclei and free electrons. Most of the positrons had been annihilated by electrons when the temperature fell below about 10 000 million degrees and became too low for electron-positron pairs to be produced from the background radiation. Neutral atoms formed when the temperature fell below about 4000 degrees. Above this temperature, the electrons in any neutral atom break free, leaving a nucleus without any electrons. When neutral atoms formed from nuclei and electrons, light could pass through the Universe without absorption, provided the clouds of neutral atoms were not too dense. At this stage, radiation now reaching us as microwave radiation set off on its long journey without interruption through space.

The Universe therefore became clear when its temperature fell below about 4000 degrees, approximately a quarter of a million times cooler than when the nuclei formed from neutrons and protons and about a quarter of a million times larger. Its size at this stage must therefore have been of the order of 1 to 10 million light years across. At this stage, it would have been about 100 000 million times older (= 250 000 × 250 000 approximately) than when nucleosynthesis started or about 100 000 years old (= 100 000 million minutes approximately). From this stage on, the Universe became darker and darker as it expanded and its matter cooled and its glow faded out. Galaxies are thought to have formed about a billion years after the Big Bang. By this stage, the Universe was probably an uneventful, dull place at a temperature of about 10 degrees above absolute zero. What happened after this to create hot spots in the Universe? What will happen to the Universe in the future? We will look in detail at these questions in the next and final chapter.

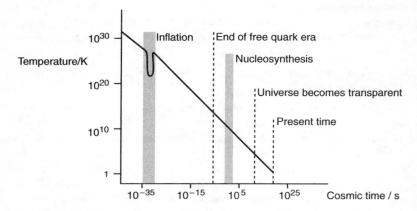

Figure 12.5   Temperature v time in the early Universe

Summary

■ **The red shift** of light from a receding galaxy is the ratio of the increase of wavelength to the emitted wavelength. The greater the red shift of a distant galaxy, the further from us and the older it is

■ **Quasars** are thought to have formed between 8 and 10 billion years ago

■ **Gamma ray bursts** are short-lived bursts of gamma radiation from sources thought to be older than quasars

■ **Inflation** of the Universe occurred when the expansion accelerated for a period of about 10^{-32} s when the Universe was about 10^{-34} s old

■ **The critical density** of the Universe is the least density that would allow the Universe to expand to infinity without collapsing back under its own gravity

■ **The density parameter** Ω is the ratio of the average density of the Universe to the critical density

■ **Baryogenesis**, the formation of quarks and antiquarks during inflation, resulted in a slight excess of quarks over antiquarks

■ **Protons and neutrons** formed from the excess quarks when the Universe was about 10 microseconds old

■ **Nucleosynthesis**, the formation of light nuclei from protons and neutrons, started to occur a minute or so after the Big Bang

Cosmic calendar

Era	Time	Temperature /K	Scale/m	Events
Pre-inflation	$10^{-43} - 10^{-35}$ s	falls to 10^{26}	10^{-51}	Temperature equalizes
Inflation	$10^{-35} - 10^{-32}$ s	supercools then recovers	expands to 10^{-1}	Baryogenesis occurs
Free quark era	$10^{-32} - 10^{-5}$ s	falls to 3×10^{12}	to 10^{13}	Quarks annihilate antiquarks
Pre-nucleosynthesis	$10^{-5} - 100$ s	falls to 10^9	to $>10^{16}$	Quarks form protons and neutrons
Nucleosynthesis	after 100 s	10^9	$>10^{16}$	Protons and neutrons join to form nuclei
Pre-atomic era	to 10^5 years	falls to 4000 K	to $>10^{22}$ m	Free nuclei and electrons
Recombination	after 10^5 years	< 4000 K	$>10^{22}$ m	Universe becomes transparent
Dark age	first few billion years	falls from 4000 K	to $>10^{25}$ m	Galaxies interact
Quasar age	after a few billion years	10 K	$>10^{25}$ m	Galaxies separating
Present	about 10^{10} years	3 K	$>10^{26}$ m	Galaxies still receding

13 INTO THE FUTURE

The search for knowledge in any branch of science generally results in new questions. Science is not about uncovering certain truths as this notion is at odds with the philosophy that a scientific theory can never be proved. No matter how many experiments are carried out to test a theory, just one experiment is sufficient to disprove it. As you approach the end of this final chapter, bear in mind the limitations of science. Based on observations, it is now accepted that the Universe originated about 10 to 15 billion years ago, and has been expanding ever since. Neutral atoms formed more or less evenly across the Universe when it was no more than a few million light years across about 100 000 years after the Big Bang. How and why did stars, galaxies and clusters and superclusters of galaxies form? Astronomers are able to see distant objects thought to be more than between 5 and 8 billion years old, corresponding to a cosmic age of a few billion years after the Big Bang. Theoreticians have been able to deduce from the laws of particle physics and relativity how the Universe developed in its first 100 000 years or so. The gap from then to between 5 and 8 billion years ago remains an unknown 'dark age' and a challenge for present-day astronomers as well as a barrier to understanding how the Universe will develop in future. In this chapter, we will look at current ideas on how the Universe evolved in its first few billion years and how the current theories predict it might develop.

The dark age of the Universe

The stars we see in the night sky are almost all part of the Milky Way galaxy, which includes the Sun in one of its spiral arms. Our home galaxy is just one of millions of millions of galaxies in all directions. Galaxies are seen to be clustered together in 'herds' up to a thousand or more strong, whichever direction we look. The number of clusters is enormous, greater than the number of people in the entire world. Galactic clusters make up superclusters, which can range up to a hundred million light years or more in size. The clusters of galaxies are seen as filament-like structures with vast empty spaces, voids, between the filaments.

The Universe is expanding, causing distant galaxies to rush away from each other. This expansion is increasing the average distance between the galaxies, but not the size of the galaxies themselves. At present, the average distance between two galactic neighbours is about five times the average size of a galaxy. Looking back in time, galaxies would have been much closer and clustering would have been less evident. Even further back, superclusters, clusters and galaxies would not have been evident at all, as little or no structure would have been evident in the tenuous dust and gas forming 100 000 years after the Big Bang.

Lumps in the primordial soup

It isn't too hard to imagine a clear thin soup, but perhaps it's a bit more difficult to imagine such a soup gradually forming lumps that become larger and larger. The image of structure developing in the Universe in a similar way is helpful up to a point, but perhaps shouldn't be taken too far, as culinary disasters of lumps of food flying apart is a bit far-fetched in a kitchen.

The density of the Universe is now of the order of 10^{-27} kilograms per cubic metre, equal approximately to one proton per cubic metre. When neutral atoms formed in the Universe, its size was of the order of a few million light years which is more than a thousand or so times smaller than its present size of the order of 10 billion light years. Its density would therefore have been about 10^{10} times (= $2000 \times 2000 \times 2000$) bigger than its present density or about 10^{-17} kilograms per cubic metre. This is much much less dense than the Earth's atmosphere at sea level, making the Universe at this stage a very very thin soup of neutral atoms, albeit at a temperature of several thousand degrees.

Let's return to our present view of the Universe. The largest superclusters are of the order of a hundred million light years across, corresponding to an angular width of about 1 degree, a little larger than the angular width of the Moon. We see the Universe as the same in all directions, provided we are looking on a scale at least 1 degree wide. Below that scale, structures such as clusters and filaments as well as voids make the Universe lumpy.

Did these structures form before the galaxies themselves formed or did the galaxies link up to form these structures?

Such questions were considered by Sir James Jeans amongst others in the first decades of the twentieth century. Jeans realized that an increase of density in a small region would attract more matter into the region, making the region more dense which would cause it to attract more matter until the region collapsed under its own gravity. Internal pressure due to the motion of the particles in the region would oppose the collapse, but Jeans worked out that collapse was inevitable if the size of the region exceeded a certain length, subsequently referred to as the Jeans length R_J. Jeans worked out that this length decreases with falling temperature but increases with falling density. The two factors combine together to cause the Jeans length to increase as the Universe expands. Although Jeans' original analysis was based on the idea of a static medium, the concept of a 'Jeans length' for gravitational collapse was developed further in later research which showed that the density variations of an expanding Universe increase as the Universe expands. If present-day galaxies were smoothed out across inter-galactic space, the density of matter would be of the order of one thousandth of the mean density of a galaxy. This follows from the average distance between two galaxies being about ten times the average size of a galaxy. As the expansion of the Universe has caused galaxies to move away from each other, it seems reasonable to suppose they were very close to each other when the Universe was ten times younger than it is now, about a billion years old. When neutral atoms formed and the Universe became transparent a hundred thousand years after the Big Bang, the Universe was a thousand times smaller. The small fluctuations of temperature detected by COBE of the order of one part in a hundred thousand, as described on page 199, represent density variations that existed when the Universe became transparent. It seems unlikely that galaxies, superclusters and voids formed in the same cosmic era. Did galaxies form before superclusters or did superclusters form before galaxies?

Evidence for structure developing in the dark age of the Universe is proving difficult to obtain. Analysis of the light from the gas surrounding some quasars reveals the presence of heavy elements, indicating star formation having taken place at redshift 4 or more. Whether or not such stars predated their host galaxies or formed from galactic clouds of dust and gas remains an open question. The fact that globular clusters of stars in the Milky Way are known to be almost as old as the Universe is evidence that structures of such size formed within one or two billion years after the Big Bang. Another open question is whether or not stars formed in these clusters or these clusters formed by drawing stars together. Computer simulations based on structure formation at constant temperature suggest a hierarchy in which smaller structures develop before larger structures.

One further piece of evidence in support of this theory is the observation that galactic clusters are less evident at high red shifts, indicating they were perhaps not a feature in the dark age. This would fit well with the theory that galaxies formed before galactic clusters but provides no clues as to whether the stars formed inside galaxies or the galaxies formed from stars.

The existence of globular clusters of stars formed over ten billion years ago does provide an indication that these globular clusters, either as discrete stars or not, formed when or before galaxies formed. If the Jeans length increased as the Universe developed, small uniform structures equivalent in mass to globular clusters could have formed first, with galaxies developing later and clusters and superclusters after that as density variations from the recombination era took effect. Clearly, much more evidence needs to be gathered before a coherent, acceptable picture is obtained of how the Universe developed during its dark age.

Galactic destroyers

Edwin Hubble classified galaxies from their appearance as either elliptical, spiral or irregular, as outlined on pages 124–5. Elliptical galaxies contain mainly red stars, ranging in number from a million or so stars in a small elliptical up to as many as a hundred million million stars in a giant elliptical. Spiral galaxies contain of the order of 100 000 million stars, including a high proportion of hot blue stars in the spiral arms and cooler red stars in the central bulge. Because red stars are known to be older than blue stars, Hubble put forward the theory that a galaxy starts out elliptical, and becomes flatter and flatter as a result of spinning faster and faster, until it develops spiral arms which are eventually thrown off to leave an irregular galaxy. This neat theory fails to explain why elliptical galaxies contain little dust in comparison with spiral galaxies and why new stars should form in the arms of a spiral galaxy, but not in an elliptical galaxy.

An alternative theory was put forward about 30 years ago by Alan Sandage, a former student of Hubble. Sandage thought that gas clouds formed in the early Universe became separate galaxies, either spiral or elliptical depending on whether the gas cloud was rotating as it collapsed under its own gravitation into a galaxy. However, observations at the Yale Observatory in America and at the Kapteyn Observatory in Holland in 1978 revealed clusters of elliptical galaxies billions of light years away containing blue galaxies, suggesting new stars in formation. More recently, observations using the Hubble Space Telescope of a remote cluster $AC_1 14$ revealed pairs of spiral galaxies on the point of merging, possibly ending up as elliptical galaxies. These observations supported computer simulations of two spiral galaxies merging, losing their spiral arms in the process and forming a single elliptical galaxy.

Very large elliptical galaxies are thought to have formed as a result of consuming smaller galaxies. Galaxies are in motion, attracting each other and tearing stars and dust off each other if they pass by close to each other. Direct collisions also occur, causing galaxies to merge and change shape. M82 shown in Plate 8 is an irregular galaxy thought to have been in collision with a nearby galaxy. The supergiant galaxy NGC 6166 in the constellation of Hercules about 400 million light years away is known to be swallowing two or three smaller galaxies. This galactic cannibal will grow larger and its increased mass will enable it to attract and destroy other neighbouring small galaxies.

The idea that elliptical galaxies formed from spiral galaxies in collision is supported by some evidence as explained above. However, other observational evidence was gathered in 1994, using the James Clerk Maxwell telescope, of a dust-filled elliptical galaxy that is billions of years old. This galaxy, 4C41.17, was observed to be at redshift 4 and found to emit infra-red radiation which could only have come from warm galactic dust. This discovery calls into question the idea that all elliptical galaxies formed as a result of spiral galaxies colliding with each other. It does seem likely that the small gas clouds formed after the Big Bang gave rise to galaxies which then formed into clusters and superclusters. More evidence is needed to decide whether or not the gas clouds developed into large dusty elliptical galaxies, which gave rise to spiral galaxies.

The quasar age a few billion years after the Big Bang suggests galactic collisions were common in that period, perhaps as a result of spiral galaxies in collision. The Milky Way and the other spiral galaxies at low redshift are perhaps galactic survivors, spared from destruction by the expansion of the Universe, taking the survivors away from each other before they could collide. However, this reprieve for the Milky Way is not permanent, as it is moving towards Andromeda for a merger in about 5 billion years hence.

Galactic ghosts

Ever since Galileo's great discovery of the moons of Jupiter, important advances in astronomy have invariably followed the inauguration of a new telescope. The Hubble Space Telescope has enabled us to see much more detail in galaxies and clusters than was possible using a ground-based telescope, no matter how powerful such a telescope might be. Before the Hubble Space Telescope, the discovery and development of photography proved to be as important to astronomers as any major new telescope. The human eye is a remarkable organ with many astonishing features. The normal human eye adapts automatically to cope with a wide range of light levels and objects over a wide range of distances. At this very moment, the eye lens in each of your eyes forms an image of the object you are observing on the retina, a light-sensitive layer at the back of the eye. The light that forms the image on the retina stimulates the cells of the retina which send nerve impulses to the brain. The pattern of impulses from the retina is interpreted by the brain, enabling it to recognize the object. When you close your eyes, the image of the object you were looking at disappears within a fraction of a second. This is how long the retinal cells take to respond to changes of light intensity. In contrast, the longer a photographic film is exposed to light, the blacker it becomes until it is completely saturated. The exposure time for an image of a dim object on a photographic film can therefore be much longer than a fraction of a second. Unlike the eye, a photographic film accumulates or 'integrates' the light, enabling the image of a very dim object to be captured.

CCD cameras are even better than ordinary cameras as the **pixels** of a CCD camera, its light-sensitive cells, are much more efficient as well as more sensitive than photographic film. In addition the CCD camera is an electronic device which can be used continuously in a remote location.

CCD cameras proved a great success on the Hubble Space Telescope, producing stunning images of objects in space never seen before in such detail. However, even before the launch of the Hubble Space Telescope, they proved their worth to astronomers in several ways, including the discovery of nearby ghostly galaxies which are too dim to register on film or to see directly. These galactic ghosts were first observed as smudges on photographs and were thought to be due to a fault caused by processing and development of the film. The advent of the CCD camera enabled astronomers from the University of Arizona to discover that these smudges are large spiral galaxies over a hundred times dimmer than normal spiral galaxies. Malin 1, the first galactic ghost to be identified, is about 800 million light years away and is almost ten times wider than the Milky Way galaxy. It would stretch 20° across the sky if it was as close as the Andromeda galaxy. But it wasn't discovered until recently, because it is much dimmer and much more spread out than normal spiral galaxies. Based on observations, astronomers now reckon there are as many ghostly galaxies as there are normal galaxies. Ghostly galaxies vary in size from dwarfs, which are much smaller than the Milky Way galaxy, to giants which are several times larger than the Milky Way galaxy. They lack massive stars and are deficient in heavy elements, suggesting they have evolved very slowly and may be billions of years old. The formation of low-intensity galaxies provides another piece in the jigsaw of how the Universe developed during its dark age although we do not know yet how the piece fits in. Astronomers do not yet know if these unexpected galaxies account for a significant amount of the estimated mass of the Universe. These galactic ghosts could prove to have an important role in the future of the Universe.

Factors for the future

The expanding Universe was predicted by Friedmann and Lemaître using Einstein's General Theory of Relativity. As we saw on page 156, Einstein introduced the cosmological constant into his model of the Universe to prevent its collapse due to gravity. Friedmann discovered an expanding Universe solution which did not need a cosmological constant. Let's look at what the expanding Universe solution tells us about the future of the Universe.

The cosmological principle

The idea of cosmic time is that the Universe looks the same from any location at the same cosmic time. The importance of this simple idea is recognized by calling it the **cosmological principle**. The distance to any distant galaxy is increasing with time and will continue to do so. The methods used to measure such distances were described in Chapter 10. A model of how the distance to any galaxy changes with time must be based on the galaxies at the same cosmic time. A false picture would be obtained if one galaxy was considered at its position a billion years ago and another galaxy was considered at its position five billion years ago.

The distance between any two galaxies may be expressed as an equation of the form

$$d(t) = R(t) \, d_0$$

where (d)t is the distance at time t after the distance was d_0, and R(t) is called the **scale factor**.

The speed of recession v(t) is the rate of change of distance which is therefore the rate of change of the scale factor, to be written R'(t), multiplied by d_0,

$$v(t) = R'(t) \, d_0$$

Combining the two equations to eliminate d_0 gives $v(t) = \dfrac{R'(t)}{R(t)} d(t)$

This may be written in the form of Hubble's Law $v(t) = H \, d(t)$ with $\dfrac{R'(t)}{R(t)}$

equal to the Hubble constant H. This analysis shows that H is a parameter that changes with time and isn't a universal constant. How the Hubble constant changes with time therefore depends on how the scale factor changes. Also, models of how the scale factor changes with time provide clues to how the Universe might develop.

Disregarding for the moment the General Theory of Relativity, the simplest model of an expanding Universe is based on the assumption that the scale factor increases steadily with time, as in Figure 13.1(a). The rate of increase of the scale factor in this model is a constant, equal to the scale factor R(t) divided by time t from zero. In other words, $R'(t) = R(t)/t$ so $R'(t)/R(t) = 1/t$. In this model, the Hubble constant $H = 1/t$, where t is the age of the Universe. As explained in Chapter 10, the present value of the Hubble constant is about 65 km s^{-1} Mpc^{-1}, giving an age of about 15 000 million years. This simple model also predicts that the Hubble constant should decrease with age which would lead to a redshift z that is not in proportion to distance. Clearly, this is not consistent with Hubble's Law, which tells us that the red shift is proportional to distance. More seriously, this simple link does not follow from Einstein's General Theory of Relativity.

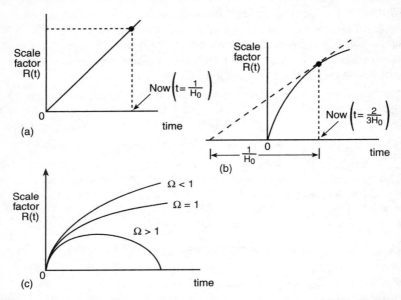

Figure 13.1 (a) Straight line expansion; (b) Prediction from Einstein's General Theory of Relativity; (c) Different futures

The Hubble constant H is the rate of change of the scale factor $R'(t)$ divided by the scale factor $R(t)$. The present value of H (which we shall write as H_0) therefore represents the rate of expansion of the Universe at the present time. As explained on page 160, a value of H equal to 100 km s^{-1} Mpc^{-1} represents an expansion factor of 0.0001 for every million years, corresponding to a 0.01% increase in the scale factor every million years or 0.1% every ten million years. Since the present value of the Hubble constant, H_0, is thought to be about 65 km s^{-1} Mpc^{-1}, the scale factor has expanded by about two-thirds of 0.1% in the last ten million years.

What does Einstein's General Theory of Relativity tell us about how the scale factor changes with time? Some of the technical details of how this great theory is applied to the expansion of the Universe are given in Appendix 4 at the end of this book. The outcome is shown in Figure 13.1(b) as a graph of the scale factor against cosmic time.

The theory predicts that the rate of change of the scale factor changes with time. Gravity is slowing the expansion at present, reducing the rate of change of the scale factor. The scale factor might continue to increase at a lesser and lesser rate or it might decrease if the Universe begins to contract rather than expand at some point in the future.

In Figure 13.1(b), the tangent to the curve at the present time intercepts the time axis at a point equal to $1/H_0$, the estimated age of the Universe without taking gravity into account. Allowing for gravity reduces the estimated age of the Universe to $2/3H_0$ which almost agrees with the estimated age of the oldest stars.

Big Crunch or Big Yawn

The expansion of the Universe might continue indefinitely or it might be reversed. Continued expansion for ever would probably be uneventful, although life would probably eventually become extinct as the temperature of the Universe would become ever closer to absolute zero. This **Big Yawn** might be preferable to a reversal of the expansion which would lead to a **Big Crunch**. The Universe would become hotter and hotter as it contracted and went through its previous stages in reverse order. Atoms and nuclei would disintegrate and turn into radiation as the Universe collapsed more and more. The weak nuclear force, the electromagnetic force and the strong nuclear force between particles would become indistinguishable as the Universe approached a cataclysmic end.

What clues do we have from the scale factor about the fate of the Universe? Figure 13.1(c) shows how the scale factor could change with time, according to general relativity. The density parameter Ω from page 203 holds the key for if the density of the Universe is greater than the critical density, the Universe faces a hot future and an eventual end in the Big Crunch. The critical density is the least density that would allow the Universe to expand to infinity. If the density of the Universe is smaller than the critical density, the Universe will expand for ever.

The density parameter Ω is defined as $\dfrac{\text{the actual density of the Universe } (\rho)}{\text{the critical density } (\rho_c)}$

■ If $\Omega > 1$, $\rho > \rho_c$ so the Universe will collapse, corresponding to a closed Universe. In this case, the curvature of space is positive, like that of a surface (e.g. a sphere) that curves back round on itself.

■ If $\Omega = 1$, $\rho = \rho_c$ so the Universe will continue to expand, eventually stopping at infinity. This scenario corresponds to a flat Universe, like a surface of zero curvature (i.e. a plane) that stretches to infinity.

■ If $\Omega < 1$, $\rho < \rho_c$ so the Universe will continue to expand without stopping, corresponding to an open Universe. The curvature of space in this case is negative, unable to stop the expansion of the Universe.

Density and dark matter

The critical density of the Universe corresponds to about half-a-dozen protons per cubic metre. The mathematics behind this statement is explained in Appendix 5 on page 245. Hydrogen and helium make up most of the luminous matter of the Universe. The density of such matter is reckoned to be about 0.01 × the critical density. This estimate is based on the measurement of the intensity of the cosmic background microwave radiation. This yields an estimate of 400 million photons per cubic metre of space. The theory of nucleosynthesis explains satisfactorily the observation that the Universe contains 25% helium to 75% hydrogen. This theory also predicts that 1 proton or neutron is produced for every 10 000 million photons. From these two facts, we can deduce that the Universe contains about 0.04 protons or neutrons per cubic metre, well below the several protons per cubic metre required for the critical density.

Galaxies are known to contain dark matter which is matter that does not emit detectable electromagnetic radiation. As we saw on page 124, the mass-to-light ratio of a galaxy is 10 to 100 times the mass-to-light ratio of the Sun, indicating the presence of dark matter 10 to 100 times more abundant than luminous matter in galaxies. In addition, galactic rotation studies indicate the presence of hidden mass in a halo that is thought to envelop the galactic disc. Another indicator of the presence of dark matter are the distorted images called 'Einstein rings' caused by light from a distant galaxy being bent by an intervening cluster of galaxies. The size of the image is a measure of the mass of the cluster causing the deflection. The results of such investigations indicate far more mass in such clusters than its luminosity suggests. Overall, the Universe is thought to contain about ten times as much dark matter as ordinary (i.e. luminous) matter. In other words, approximately 90% of the mass of the Universe is in the form of dark matter. If this is the case, the overall density of matter in the Universe is ten times greater than the density of ordinary matter, corresponding to about 0.4 protons per cubic metre which is still well below the critical density. On the basis of evidence thus far, the Universe would therefore seem to be open, expanding for ever and heading for the Big Yawn!

MACHOS and WIMPS

What form could the dark matter in the galaxies be in? The standard model of the Big Bang predicts the ratio of photons to protons and neutrons (page 188). Astronomers have ruled out the possible ways in which large numbers of protons and neutrons could be hidden away in space. The dark matter cannot be cold dust, debris or black dwarfs thrown into space by supernovae long ago, as this debris would form new stars with an abundance of heavy elements. New stars in the arms of spiral galaxies do not contain heavy elements in abundance. Low-mass stars a little larger than Jupiter would be very faint and could account for the missing matter, but a large number of such stars would result in a missing population of larger stars bright enough to see, but smaller than the smallest stars ever observed. Black holes cannot account for the missing mass, as small black holes would have evaporated long ago, and there is no evidence for large numbers of large black holes. The dark matter could be due to massive compact objects in some unknown form in the haloes of galaxies. Although there is no observational evidence for such objects, astronomers refer to these hypothetical objects as MACHOs, an abbreviation for MAssive Compact Halo Objects. Other imaginative possibilities to account for dark matter have emerged from particle physics theory, including WIMPs or Weakly Interacting Massive Particles, and other hypothetical particles, all collectively known as cold dark matter, that might have formed in the cauldron of the Big Bang. Astronomers, and their theoretical counterparts, seem to have been driven to the outer reaches of their ideas because dark matter constitutes over 90% of the mass of the Universe but its location is a mystery.

Most of the matter in the Universe is dark matter in the form of particles which are not protons or neutrons. Vast clouds of electrons would be unstable and would interact with light. So what other types of known particles could make up dark matter? One possible candidate is the neutrino and its antimatter counterpart, the antineutrino. We met these particles on page 179. At this very moment, billions of neutrinos are passing through your body from the Sun, streaming into space at or near the speed of light. They are produced in the process of beta decay when a proton changes into a neutron or *vice versa*. Neutrinos are reckoned to have been released as free particles when the Universe was aged about one second. Calculations indicate about 1 neutrino was released for every 4 photons released, giving a neutrino density of about 100 million per cubic metre of space. To make up the missing mass of the Universe which is equivalent to about 0.5 protons per cubic metre, the mass of a neutrino would need to be no less than about 5 billionths (= 0.5/100 million) of the mass of a proton. As explained on page 185, the rest mass of a proton is about 1000 MeV. Thus the mass of a neutrino would need to be about 5 eV if all the known dark matter is to be accounted for by neutrinos. At the present time, the mass of the neutrino remains uncertain and is the subject of considerable experimental and theoretical research. If the mass of the neutrino turns out to be more than about 50 eV, which is ten times greater than needed to account for all the known dark matter, the Universe could be in for a very sticky end in the Big Crunch! One further twist is that galaxy formation calculations favour cold dark matter, not neutrinos.

An uncertain future

Modern science provides us with a model of the Universe which few would believe if it was not supported by well-accepted scientific evidence. The earth-centred Universe of ancient philosophers might seem more acceptable than the astonishing modern theory that the Universe originated from a point that was created over ten billion years ago. As explained on page 170, the observation of background microwave radiation from all directions in space provided the crucial evidence in support of the Big Bang theory. Before the discovery of this background microwave radiation, **the Steady State theory of the Universe** provided a respectable alternative model of the Universe. This theory was put forward in the 1950s by three prominent theoretical astronomers, Fred Hoyle, Hermann Bondi and Thomas Gold. At that time, the accepted value of the Hubble constant made the age of the Universe younger than the Earth, so the Big Bang theory did not find favour throughout the scientific community. The Steady State theory was based on the idea that the Universe has been expanding for ever and matter is being continuously created in the voids between galaxies as they move apart. The theory assumes the Universe appears the same on a cosmological scale at all points in space and time. The Hubble constant is a true constant in this theory, the same for all time. The two theories provided a source of controversy for some years until the discovery of background microwave radiation. This was explained as a consequence of the Big Bang and has no place in the Steady State theory.

The history of science tells us never to sign up once and for all to a theory. We can never know how much science remains undiscovered, and we can not be sure that a given theory is the last word. Just one reliable experiment is all that is needed to dispense with, or to require a new look at, a theory. The Big Bang theory, like any scientific theory, is there to be probed and tested. An alternative explanation for Hubble's Law is based on the idea that galactic redshift decreases with time rather than distance: a distant galaxy has a large redshift because it was younger when it emitted its light according to this idea. However, this theory needs a leap of faith since it assumes objects gain mass and become heavier as they age – this may be true for us humans, but there is no evidence of this effect for galaxies.

Big questions still surround the Big Bang theory. For example, how did the energy of the Universe originate? This question is simple enough but the answer is not yet known. One of the key principles of modern science is that energy (including energy in the form of mass) cannot be created or destroyed. This principle has successfully withstood repeated experimental tests over almost two centuries. Why should it be ruled out from the origin of the Universe and taken to be valid only from some moment, however soon, after the Big Bang?

The one-off nature of the Big Bang is under challenge as a result of attempts to understand these fundamental questions. The answers may lie in new discoveries in high-energy particle physics, being linked up to quantum theory and a newly emerging branch of science known as chaos theory. Chapter 11 touched on the Large Hadron Collider at CERN in Geneva. This colossal accelerator is designed to smash particles together with energies forty times greater than is possible at present. With this machine, particle physicists will find out if a predicted particle called the Higgs boson actually exists. The standard model of particle physics successfully explains the early stages of the Big Bang, and makes a further prediction that mass is due to particles coupling to Higgs bosons which are the quanta of an energy field that permeates space. If the Higgs boson is discovered, particle physicists and cosmologists will breathe a deep sigh of relief that the standard model and the Big Bang remain intact. However, if the Higgs boson is not discovered, these theories will be called into question. The discovery of the Higgs boson would provide vital clues about the nature and properties of mass which will help to understand the creation of matter in the Big Bang.

Universal chaos

According to Isaac Newton, the Universe is like a giant clockwork machine in which all its parts behave predictably in accordance with the laws of science, notably Newton's laws. These classical laws leave little room for freedom as they predict the behaviour of any mechanical system, given a complete specification of its condition at the outset. The planets move round the Sun on paths which are very predictable, enabling space missions to be planned and carried out successfully. If the planets moved erratically, space missions would be hit-and-miss events, more often misses than hits. Newton's laws remain satisfactory for working out the behaviour of any mechanical system except when quantum mechanics is needed (e.g. on a molecular or sub-molecular scale or at very low temperatures) or when relativity theory is necessary (e.g. at speeds approaching the speed of light). It came as a great surprise when about 20 years ago, systems were discovered that were not always predictable in accordance with Newton's laws, and that displayed chaotic behaviour under certain conditions. These discoveries opened up a new branch of science which has become known as **chaos theory**. The mathematics of chaos theory works out in such a way that it is possible for a tiny disturbance in part of a system to have a disastrous effect. The discoveries in chaos theory have resulted from the enormous increase in computing power that has become available in recent decades. Quantum fluctuations in the early pre-inflation Universe may well have expanded a small part of the Universe beyond a point of no-return beyond which further expansion became unavoidable. Some theoreticians reckon that new universes are being created by inflation at points in our Universe!

From chaos to expansion

Appendix 6 is a simple spreadsheet program for a one-dimensional system which changes its position according to how far it is from each of two fixed points. Its evolution depends on its initial position and can change very dramatically from chaos to expansion with just a small change of its initial position, as shown in Figure 13.2.

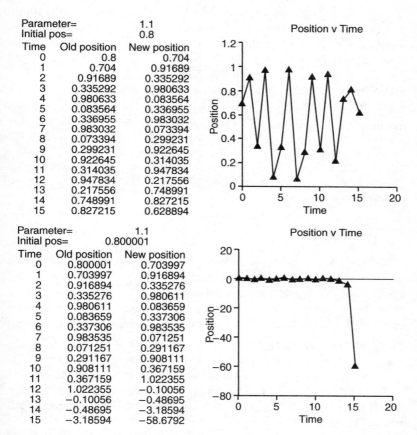

Parameter=	1.1	
Initial pos=	0.8	
Time	Old position	New position
0	0.8	0.704
1	0.704	0.91689
2	0.91689	0.335292
3	0.335292	0.980633
4	0.980633	0.083564
5	0.083564	0.336955
6	0.336955	0.983032
7	0.983032	0.073394
8	0.073394	0.299231
9	0.299231	0.922645
10	0.922645	0.314035
11	0.314035	0.947834
12	0.947834	0.217556
13	0.217556	0.748991
14	0.748991	0.827215
15	0.827215	0.628894

Parameter=	1.1	
Initial pos=	0.800001	
Time	Old position	New position
0	0.800001	0.703997
1	0.703997	0.916894
2	0.916894	0.335276
3	0.335276	0.980611
4	0.980611	0.083659
5	0.083659	0.337306
6	0.337306	0.983535
7	0.983535	0.071251
8	0.071251	0.291167
9	0.291167	0.908111
10	0.908111	0.367159
11	0.367159	1.022355
12	1.022355	−0.10056
13	−0.10056	−0.48695
14	−0.48695	−3.18594
15	−3.18594	−58.6792

Figure 13.2   Out of chaos

Beyond science

We have looked in some detail at the strong scientific evidence in support of the theory that the Universe is expanding from a point of creation over ten billion years ago. Without the weight of scientific evidence behind it, this theory would be little better than Ptolemy's geocentric model of the Universe. If you ask a friend unaware of the Big Bang theory about the relative merits of the geocentric model and the Big Bang theory, you are as likely to find support for the geocentric model as for the Big Bang. Yet the weight of scientific evidence is unambiguously behind the Big Bang theory. Modern science has provided answers to some of the big questions in cosmology. However, many questions remain and many more questions have been raised by the Big Bang theory. The observations and experiments carried out by scientists and astronomers provide essential guidance for the theoreticians. Before the Scientific Age was ushered in by Copernicus and Galileo, few dared to challenge the medieval Church on the geocentric model of the Universe. The freedoms we now take for granted (such as human rights and democracy) would not perhaps have been won if the struggle for intellectual freedom, the hallmark of modern science since Galileo, had been lost. Like science, democracy seems an untidy sphere of human activity with no certain answers. Let's hope no certain answers are ever found.

Summary

The density parameter Ω is defined as $\dfrac{\text{the actual density of the Universe } \rho}{\text{the critical density } \rho c}$

- If $\Omega > 1$, $\rho > \rho_c$: the Universe will collapse
- If $\Omega = 1$, $\rho = \rho_c$: the Universe will continue to expand
- If $\Omega < 1$, $\rho < \rho_c$: the Universe will continue to expand without stopping

Dark matter constitutes at least 90% of the matter of the Universe. More hidden mass may exist in objects or particles not known at present. If neutrinos are not massless, they could add significantly to the total mass of the Universe. If the rest mass of the neutrino exceeds about 50 eV, the density parameter would be greater than 1 and the Universe would then ultimately collapse.

GLOSSARY

Absolute magnitude M   The magnitude of a star if it was at a distance of 10 parsecs from us.

Antimatter   For every type of particle, there is an antiparticle which has opposite charge, the same mass and will annihilate its particle counterpart and itself to form high-energy radiation. High-energy radiation is capable of producing particle-antiparticle pairs. The positron is the antimatter counterpart of the electron.

Astronomical unit (AU)   The mean distance from the Sun to the Earth, equal to 150 million kilometres.

Atomic structure   Every atom contains a nucleus which consists of protons and neutrons. The space surrounding the nucleus is occupied by electrons. The hydrogen atom consisting of 1 proton and 1 electron is the simplest atom.

Baryogenesis   The formation of quarks and antiquarks during inflation; resulted in a slight excess of quarks over antiquarks.

Bending of light by gravity
1 Light grazing the Sun from a star as it is occulted by the Sun during a solar eclipse.
2 Double images of quasars due to gravitational lensing.

Binary star system   Two or more stars in orbit about each other. Applying Kepler's 3rd Law to binaries gives
Total mass $\times$ (Time period)^2 = Separation^3.

Black hole   A massive object which light cannot escape from.

Black hole rules
1 **The Schwarzschild radius** defines a sphere surrounding a black hole which no object or light can escape from.
2 **The event horizon** of a black hole is the surface of the Schwarzschild sphere surrounding a black hole.

3 Thermal radiation from a black hole due to particle-antiparticle pairs created just outside the event horizon causes a black hole to evaporate. The rate of evaporation of a black hole is insignificant unless its mass is less than 10^{12} kg.

Boson The force-carrier of the weak nuclear force.

Cepheid variables Variable stars with variability periods from days to weeks.

Circumpolar stars Stars that never set and can be seen at any time of year on a clear night.

Classification of stars O = 30 000; B = 20 000; A = 10 000; F = 8000; G = 6000; K = 4000; M = 3000 degrees.

Clusters Galaxies are in clusters which are often in clusters of clusters (superclusters) with voids between. 90% of the mass of a galaxy may be dark matter. Clusters are distributed evenly over distances greater than about 100 million parsecs.

Comet An object in elliptical orbit round the Sun, stretching far beyond the orbit of Pluto.

Constellation A pattern of stars in the night sky. The night sky is mapped out in 88 constellations.

Copernican system The Sun not the Earth is at the centre of the system of planets. The planets including the Earth orbit the Sun.

Critical density The least density that would allow the Universe to expand to infinity without collapsing back under its own gravity.

Dark matter This constitutes at least 90% of the matter of the Universe. More hidden mass may exist in objects or particles not known at present. If neutrinos are not massless, they could add significantly to the total mass of the Universe. If the rest mass of the neutrino exceeds about 50 eV, the density parameter would be greater than 1 and the Universe would then ultimately collapse.

Declination The angle between the star and the Celestial Equator.

Degree 1 degree = 60 minutes of arc = 3600 seconds of arc.

Density parameter Ω This is the ratio of the average density of the Universe to the critical density.

Diffraction The spreading of waves round an obstacle or through a gap.

Distance in parsecs to a star = 1/parallax angle in seconds of arc.

Doppler effect   The change of wavelength of light from a star due to the speed at which it is approaching or receding. The speed is given by (change of wavelength/source wavelength) × speed of light.

Ecliptic   The apparent path of the Sun through the constellations. It is the Earth's orbit projected on the Celestial Sphere.

Einstein's mass energy equation $E = mc^2$ where m is the mass change of an object if its energy changes by E.

Einstein's Special Theory of Relativity
1   The speed of light through space is always the same, regardless of the motion of the source or the observer.
2   The laws of physics in the form of equations should be the same in any inertial frame of reference.

Electromagnetic forces   act between charged objects due to the exchange of virtual photons, repel if the charges are like, attract if unlike, become weaker with distance, and have infinite range.

Electromagnetic waves (in order of increasing frequency)   Radio waves, microwaves, infra-red radiation, light, ultraviolet radiation, X and gamma radiation. All electromagnetic waves travel at a speed of 300 000 km/s through free space.

Electron   A negative particle found in every atom.

Equinoxes   Mid-Spring and mid-Autumn, when daylight and darkness are of equal duration. The equinoxes correspond to the two points of intersection of the ecliptic and the Celestial Equator.

Evolution of a star   The stages a star goes through from formation until it emits no more light.

First Point of Aries   This is the point of intersection corresponding to the vernal (i.e. spring equinox).

Galaxy   An assembly of millions of millions of stars.

Gamma ray burst   A short-lived burst of gamma radiation from a distant source thought to be older than quasars.

Gluon   The force-carrier responsible for holding quarks together in threes as protons or neutrons.

Gravitational forces   act between objects due to their mass, are attractive only, become weaker with distance and have infinite range.

Gravitational red shift   is the decrease of the frequency of a photon escaping from the gravitational field of a massive body.

Helium 4   An isotope of helium with 2 protons and 2 neutrons in every atomic nucleus.

Hertzsprung–Russell diagram   A graph of absolute magnitude against colour or temperature for the stars.

Higgs boson   The quantum of the scalar field that causes elementary particles to have mass.

Hubble's classification of galaxies   S(B) a–d for spiral galaxies (with B, if the centre is bar-shaped); E0–E7 for elliptical galaxies (see page 124).

Hubble's Law   The speed of recession of a distant galaxy = Hd, where d is its distance and H is the Hubble constant. The value of H is thought to be about 65 km s^{-1} Mpc^{-1} or 65/1 000 000 Myr^{-1}.

Hydrogen   Hydrogen atoms in space emit radio waves at 21 cm mostly from dust and gas in spiral arms.

Inflation   This is thought to have occurred when the expansion of the Universe accelerated for a period of about 10^{-32} s when the Universe was about 10^{-34} s old.

Inverse square law of
  force   See Newton's law of gravitation.
  radiation   The intensity of light from a point source is inversely proportional to the square of its distance.

Ionisation   The process of an uncharged atom becoming charged.

Isotopes   Atoms of an element with different masses due to differing numbers of neutrons in the atomic nucleus.

Jeans length   Diameter of a cloud of dust or gas in a dust-filled Universe that would collapse due to gravity.

Kepler's laws of planetary motion
1 Each planet moves round the Sun on an elliptical orbit.
2 The speed of each planet changes as it moves along its orbit. The progress of an imaginary line from the Sun to each planet varies as the inverse of the square of the distance from the planet to the Sun.
3 The square of the period of a planet is proportional to the cube of its mean distance from the Sun.

Large Hadron Collider   The 7000 GeV particle accelerator at CERN.

Lepton   An elementary particle that is not a quark.

Light year The distance travelled by light in one year. The nearest star is about 4 light years away. The most distant galaxies are thought to be over 6000 million light years away.

Luminosity or power of a star The energy per second radiated by a star. The luminosity of a Main Sequence star is approximately proportional to the cube of its mass.

Magnitude A scale of star brightness where magnitude 6 can just be seen unaided. Every 5 magnitudes extra corresponds to one hundred times less light received.

Mass The gravitational mass of an object (i.e. its mass measured by weighing it) is equal to its inertial mass (its mass measured by changing its velocity).

Matter and antimatter Matter exists as either quarks or leptons. A proton consists of two up quarks and a down quark. A neutron consists of two down quarks and an up quark. The lepton family consists of electrons and neutrinos. Antimatter consists of antiquarks and antileptons.

Meson A particle lighter than a proton.

Microwave background radiation Microwaves from all directions in space left over from the Big Bang.

Milky Way galaxy The Sun lies in the Orion arm of the Milky Way galaxy. The galactic centre lies beyond the Sagittarius arm of the Milky way galaxy.

Models of the Universe
1 **Newton's Universe** is static and infinite.
2 **Einstein's static Universe** requires the introduction of a cosmological force of repulsion.
3 **Friedmann's Universe** expands without the necessity of a cosmological force.
4 **The Steady State model** supposes that matter is continuously created as matter spreads out.
5 **The Big Bang model** supposes that the Universe was created in an explosion at a point billions of years ago, and has been expanding ever since.

Neutrino   A neutral particle emitted when a beta particle is emitted by an unstable nucleus.

Neutron   A neutral particle in every atomic nucleus except the lightest (and most common) form of hydrogen.

Newton's law of gravitation   For two masses m_1 and m_2 spaced apart at distance r between their centres of gravity, a force F of gravitational attraction exists between any two objects given by $F = G \, m_1 m_2 / r^2$ where G is the Universal Constant of Gravitation.

Nucleosynthesis   The formation of light nuclei from protons and neutrons, said to have occurred a minute or so after the Big Bang.

Olbers' Paradox   The night sky is dark because the Universe is finite and expanding.

Opposition   This is when a star or planet is in the opposite direction to the Sun.

Parallax   The change of apparent position of a nearby object relative to distant objects due to the movement of the observer.

Parallax angle of a star   The angle between the line from the star to the Sun and the line from the star to the Earth (when the Earth–Sun line is at 90° to the star–Sun line).

Parsec   The distance to a star that makes an angle of exactly one second of arc with the Sun and the Earth.

Photon   A photon is a quantum of electromagnetic energy in the form of a packet of electromagnetic waves. The energy of a photon is in proportion to the frequency of the waves.

Pion   A quark-antiquark pair exchanged when protons or neutrons interact through the strong nuclear force.

Planet   A large object in orbit about a star. The Sun's nine planets are Mercury, Venus, Earth, Mars, Jupiter, Saturn, Uranus, Neptune and Pluto. The planets are visible from Earth because they reflect sunlight.

Population I stars   Hot, blue, metal-rich and young, to be found in the spiral arms of galaxies.

Population II stars   Red giants, metal-deficient and old, to be found in globular clusters and in the galactic centre.

Positron   The antimatter counterpart of the electron.

Principle of Equivalence   This asserts that the effects of gravity and of accelerated motion are identical.

Proper motion   The movement of a star across the line of sight.

Proton   A positive particle in every atomic nucleus. The proton is the nucleus of the hydrogen atom.

Protostar   A star in formation.

Quark   An elementary particle of which protons and neutrons are composed. Every proton is composed of two up quarks and a down quark. Every neutron is composed of one up and two down quarks.

Quasar   A quasar is thought to be a distant active galactic nucleus where a massive black hole is attracting surrounding matter. Quasars are thought to have formed between 8 and 10 billion years ago.

Radian   A measure of angle where 360 degrees equals 2π radians.

Radioactivity   A radioactive substance consists of atoms with unstable nuclei. An unstable nucleus becomes stable by emitting either an alpha particle (which is composed of two protons and two neutrons) or a beta particle (which is an electron created and emitted in the nucleus when a neutron changes to a proton) or a gamma photon.

Red giant   A K or M class star that is much larger than the Sun.

Redshift   of light from a receding galaxy. This is the ratio of the increase of wavelength to the emitted wavelength. The greater the red shift of a distant galaxy, the further from us and the older it is.

Refraction   The change of direction of light on passing from one transparent medium to another.

Resolving power   The minimum angular separation between two point objects that can just be resolved.

Retrograde motion   This occurs when the East-to-West progress of an outer planet through the constellations reverses for a period of weeks before resuming its normal progress.

Right ascension of a star   The angle along the Celestial Equator from the Great Circle through the star to the First Point of Aries.

Spectra

1 A continuous spectrum includes all the colours of the spectrum.

2 A line emission spectrum consists of coloured lines at definite wavelengths.

3 A line absorption spectrum consists of black lines at definite wavelengths on a continuous spectrum.

Spectroscopic binary   A binary system that cannot be resolved into two stars by observation. The lines of its spectrum repeatedly open and close.

Strong nuclear forces   act between neutrons and protons due to the exchange of pions, have a range of about 10^{-15} metres, are attractive except at very short range and act equally between neutrons and protons.

Supercluster   A cluster of clusters of galaxies.

Supernova   A supernova is the explosion of a white dwarf after it collapses. This happens if the mass of the white dwarf exceeds 1.4 solar masses. Heavy elements are formed in a supernova explosion.

Void   Empty space between superclusters.

Wavelength   The distance from one peak of a wave to the next.

Weak nuclear forces   are responsible for beta decay processes in which a boson is created and decays into a lepton and an antilepton over a range of no more than about 10^{-18} m.

White dwarf   An O or B class star that is much smaller than the Sun.

APPENDIX 1
PROOF OF KEPLER's 3rd LAW

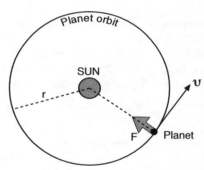

Figure A1

The diagram represents a planet of mass m on a circular orbit of radius r around the Sun. Newton showed that any object moving round a circle at steady speed υ must be acted on by a force F towards the centre of the circle given by the formula $F = \dfrac{m\upsilon^2}{r}$

For a planet in orbit round the Sun, this force is due to the gravitational attraction between the planet and the Sun. Newton's law of gravitation gives this force as $F = G\dfrac{Mm}{r^2}$ where M represents the mass of the Sun.

Therefore $G\dfrac{Mm}{r^2} = \dfrac{m\upsilon^2}{r}$

Cancelling m on both sides and rearranging gives $\upsilon^2 = \dfrac{GM}{r}$

Since speed $\upsilon = \dfrac{\text{circumference}}{\text{time period}} = \dfrac{2\pi r}{T}$, then substituting for υ gives $\dfrac{(2\pi r)^2}{T^2} = \dfrac{GM}{r}$

Rearranging this equation gives $T^2 = kr^3$ where $k = \dfrac{4\pi^2}{GM}$, in agreement with Kepler's 3rd Law.

APPENDIX 2
MORE ABOUT
SPECIAL RELATIVITY

Consider the equation for the position of a light wave spreading out from a point source. If an observer who is stationary relative to the light source could see the wave spreading out, passing distance markers as it progresses, he or she would observe its distance from the point source increasing with time, according to the equation

Distance moved, r = speed of light, c × time, t

To keep the analysis simple, imagine the distance markers are in two perpendicular directions only, East and North. We will use x for one of these distances (East) and y for the other (North). Using Pythagoras' right-angle triangle rule $r^2 = x^2 + y^2$

$$\therefore x^2 + y^2 = c^2t^2 \tag{1}$$

A second observer moving at speed V relative to the first observer in the x-direction would make measurements using his or her own coordinate system to arrive at a similar equation.

Distance moved, r′ = speed of light, c × time, t

where r′ and t′ are measurements made by the second observer.

Using x′ and y′ for this observer's two-coordinate system therefore gives equation 2

$$x'^2 + y'^2 = c^2t'^2 \tag{2}$$

If the two observers started off at the same point, according to Newton's ideas of space and time, then $x' = x - Vt, \quad y' = y \quad$ and $\quad t' = t$

Substituting the expressions for x′, y′ and t′ into equation 2 will not give equation 1. Using Newton's ideas, it's not possible to transform one equation into the other. What Einstein did was to derive a set of transformation formulas that do work when the coordinate system is switched. For the record, they are given below but don't take fright, as we only look at what they mean!

$$x' = \gamma\,(x - Vt)$$

$y' = y$ (the y-coordinate isn't affected by motion in the x-direction)

$$t' = \gamma\left(t - \frac{Vx}{c^2}\right) \quad \text{where} \quad \gamma = \frac{1}{\sqrt{\left(1 - \frac{V^2}{c^2}\right)}}$$

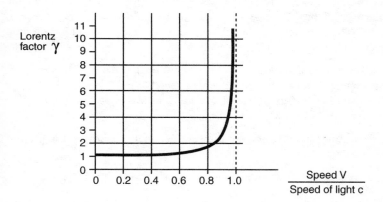

Figure A2   The Lorentz factor

The Lorentz factor is the key to why nothing can travel faster than light. Figure A2 shows how γ changes with speed V. It increases from 1 at zero speed and it becomes infinitely large at $V = c$.

APPENDIX 3
TAKING ACCOUNT
OF GRAVITY

Gravity is an attractive force and it is not unreasonable to suppose that the expansion of the Universe is or was partly held back by the force of gravity. The expanding galaxies lose kinetic energy and gain potential energy as they move away from each other.

- The kinetic energy of an object is proportional to the square of its speed υ, at speeds much less than the speed of light.
- The potential energy of an object on the surface of a sphere is proportional to l/r, where r is the sphere's radius.

For a receding galaxy, assume the loss of kinetic energy = the gain of potential energy.

Therefore υ^2 is proportional to $\dfrac{l}{r}$ which makes υ proportional to $\dfrac{1}{r^{1/2}}$

Hence $\upsilon = \dfrac{dr}{dt} = kr^{-1/2}$ where k is the constant of proportionality.

Integrating this expression therefore gives $2r^{3/2} = 3kt$, where t represents time.

Substituting $k = \upsilon r^{1/2}$ (from $\upsilon = kr^{-1/2}$ rearranged) therefore gives

$$2r = 3\upsilon t$$

hence $\qquad \upsilon = \dfrac{2r}{3t}$

The Hubble constant H according to this relationship is therefore $\dfrac{2}{3t}$

Hence $t = \dfrac{2}{3H}$

APPENDIX 4
NOTES ON EINSTEIN'S GENERAL THEORY OF RELATIVITY

The curvature of space is defined by the tensor $R_{ij} - \frac{1}{2}g_{ij}R$ which is known as the Einstein tensor E_{ij}. The tensor R_{ij} is a complicated function of the metric tensor and its derivatives. Einstein showed that the curvature of space as expressed through the Einstein tensor E_{ij} is related to the distribution of matter as expressed by means of the mass-energy tensor T_{ij} according to the equation

$$E_{ij} = - \frac{8\pi\,G}{c^2}\,T_{ij}$$

where c is the speed of light in free space and G is the universal constant of gravitation.

1 For empty space surrounding a spherical mass M, the equation becomes $E_{ij} = 0$. The solution of this equation was first worked out by Karl Schwarzschild who showed that light cannot escape from a light source near M if the source is closer than a distance equal to
$$\frac{2GM}{c^2}$$

2 For a Universe consisting of gas of density ρ, Einstein showed that the full equation relating the curvature of space to the distribution of matter could be simplified to an equation for the scale factor R and its rate of change R′, as follows:

$$R'^2 = \frac{8\pi G\rho}{3}\,R^2 - Kc^2 + \frac{1}{3}\lambda R^2$$

where K represents the curvature of the Universe and λ is the so-called cosmological constant. As explained on page 155, Einstein introduced this constant to prevent a solution in which the Universe collapses, and this assumption proved to be unnecessary when an 'expanding Universe' solution was discovered without λ.

Assuming $\lambda = 0$ then $R'^2 = \dfrac{8\pi G \rho}{3} R^2 - Kc^2$

■ Using the subscript '0' to indicate present-day values gives
$R'_0{}^2 = \dfrac{8\pi G \rho_0 R_0{}^2 - Kc^2}{3}$

Let $\rho_c = \dfrac{3H_0{}^2}{8\pi G}$ so $H^2_0 = \dfrac{8\pi G \rho_c}{3} = \dfrac{8\pi G \rho_0}{3\Omega}$

where the density parameter $\Omega = \rho_0/\rho_c$

Therefore $R'_0{}^2 = \Omega H_0{}^2 R_0{}^2 - Kc^2$

Using Hubble's Law in the form $R'_0 = H_0 R_0$

then we see that $K = H_0{}^2 R_0{}^2 (\Omega - 1)/c^2$

■ Also, since the mass of the Universe is more or less constant now, its density $\rho =$ its present-day density $\rho_0 \times R_0{}^3/R^3$, where R_0 is the present-day scale factor

Hence $R'^2 = \dfrac{8\pi G \rho_0 R_0{}^2}{3R} - Kc^2$

■ If $\Omega > 1$ then K is positive so the curvature of the Universe is spherical. As it expands, its rate of expansion decreases to zero when $\dfrac{8\pi G \rho_0 R_0{}^2}{3R} = Kc^2$ and it then collapses.

■ If $\Omega = 1$ then K is zero so the Universe is flat. As it expands, its rate of expansion decreases to zero at infinity (as R becomes infinite) after an infinite time.

■ If $\Omega < 1$ then K is negative so the curvature of the Universe is hyperbolic (i.e. open). As it expands, its rate of expansion does not decrease to zero and it continues to expand for ever.

Note that for $\Omega = 1$, then $\rho_0 = \rho_c$ where $\rho_c = \dfrac{3 H_0{}^2}{8\pi G}$.

Since the density parameter determines whether or not the Universe collapses or expands in future, ρ_c is referred to as the **critical density**. If the present-day density exceeds this value, the Universe will collapse.

APPENDIX 5
CALCULATION OF THE CRITICAL DENSITY OF THE UNIVERSE

For a flat Universe, general relativity predicts that the rate of change of the scale factor R' depends on the density of the Universe in accordance with the equation

$$R'^2 = \frac{8\pi G\rho}{3} R^2$$

Note This equation can also be derived using Newton's laws by equating the gravitational potential energy of a galaxy of mass m 'at the edge of the Universe' to its kinetic energy at speed υ,

$$\frac{GMm}{r} = \tfrac{1}{2} m\upsilon^2$$

where r is the distance to the edge of the universe and M is the mass of the Universe.

$$\therefore \upsilon^2 = \frac{2 \, GM}{r}$$

Since mass M = volume × density and the volume of a sphere of radius $r = \tfrac{4}{3}\pi r^3$, the equation for R' follows if R' is substituted for υ and R is substituted for r.

As explained in Appendix 4, the critical density ρ_c is defined from this equation

$$\rho_c = \frac{3H_0{}^2}{8\pi G} \, , \text{ where } H_0 \text{ is the present value of the Hubble constant.}$$

Substituting $H_0 = 2.1 \times 10^{-18}$ s^{-1} (= 65 km s^{-1} Mpc^{-1} = 65/1 000 000 Myr^{-1} converted to s^{-1}) and $G = 6.7 \times 10^{-11}$ N m^2 kg^{-2} gives $\rho_c = 7.9 \times 10^{-27}$ kg m^{-3}. This corresponds to about 5 protons per cubic metre!

APPENDIX 6
SPREADSHEET FOR 'OUT OF CHAOS' FIGURE 13.2

Note The symbol $ is for absolute cell references

1 Key into cells A1 and A2, the text expressions 'Parameter =' and 'Initial position ='.

2 Key the parameter value into cell B1. Key the initial position into cell B2.

3 Key text headings 'Time', 'Old position' and 'New position' into cells A3, B3 and C3 respectively.

4 Key 0 into cell A4; B2 into cell B4 and the formula 4*B1*D2*(1–(D2)) into cell C4 for the new position calculated from the old position and the parameter according to the equation *new position = 4 × parameter value × old position × (1 – old position)*.

5 Key (A4) +1 into cell A5; key C4 into cell B5; copy the contents of C4 into C5.

6 Copy cells A5, B5 and C5 down columns A, B and C.

7 To chart the results, plot col. C on the y-axis against col. A on the x-axis.

INDEX

absolute magnitude M 70
age of a star 105
age of the Universe 158
alpha particle 58
antimatter 59, 187
antiparticle 59
antiquark 208
astronomical unit (AU) 11
atom 58

baryogenesis 208
bending of light by gravity
    double images of quasars 141
    light grazing the Sun 150
beta particle 58
binary star system 85
black hole 133
blue shift 87
boson 182

Celestial Equator 17
Celestial Sphere 16
Cepheid variable 110
Chandresakhar limit 101
Chandresakhar, Subrahmanyan 101
circumpolar stars 18
classification of stars 81
clusters 121
comet 9
constellation 7
Copernican system 28
cosmic time 194
cosmological constant 155
cosmological principle 220
critical density 201

dark matter 125, 225
declination 22
density parameter 203
distance modulus 71
distance to a star 64
Doppler effect 81

ecliptic 19
Eddington, Sir Arthur 150
Einstein 53
Einstein's mass energy equation 57
Einstein's Special Theory of Relativity 54
electromagnetic forces 177
electromagnetic waves 49
electron 57
electron volt (eV) 85
elliptical galaxies 124
equinoxes 20
event horizon 144
Evidence for the Big Bang Theory
    Hubble's Law 171
    Microwave background radiation 170
    3-to-1 hydrogen to helium 170
evolution of a star 96
expansion of the Universe 131

Faraday, Michael 175
Feynman, Richard 177
First Point of Aries 22

galactic centre 121
galaxies 123
Galileo, Galilei 34–8
gamma photon 58
gamma ray burst 197
General relativity predictions
    black holes   145
    deflection of light by gravity 150
    gravitational red shift 151
    gravitational waves 157
    precession of the orbit of a planet 149
gluon 182
gravitational forces 40
gravitational lensing 141
gravitational red shift 142
Guth, Alan 201

Hawking, Stephen 145
heavy elements 100
helium 93

Hertzsprung–Russell diagram 74
Higgs boson 228
Hubble constant 130
Hubble, Edwin 132
Hubble's classification of galaxies 123
Hubble's Law 130
hydrogen 93

inertial frame of reference 54
inflation 201
infra-red radiation 49
inverse square law 40, 71

Jeans length 216
Jeans, Sir James 169

Kepler, Johann 29
Kepler's laws of planetary motion 32

Large Hadron Collider 185
lepton 184
light year 12
luminosity 90

magnitude 67
Main Sequence star 74
mass 138
matter 187
meson 180
microwave background radiation 170
microwaves 49
Milky Way galaxy 13
minute of arc 65
models of the Universe
    Big Bang Model 161
    Einstein's static Universe
    Friedmann's Universe 156
    Newton's Universe 154
    Steady State model 169

neutrino 179
neutron 58
neutron star 102
Newton, Sir Isaac 38
Newton's law of gravitation 40
nova 99
nucleosynthesis 210
nucleus 58

Olbers' Paradox 155
opposition 25

parallax 64
parallax angle 65
parsec 65
photon 51

pion 182
planet 8
Population I stars 115
Population II stars 115
positron 59
power 91
Principle of Equivalence 138
proper motion 66
proton 58
protostar 97
Ptolemy 25
pulsar 103

quantum 50
quark 182
quasar 133, 193

radioactivity 58
radio waves
    in the electromagnetic spectrum 49
    at 21 cm wavelength 119
red giant 75
red shift 129
retrograde motion 25
right ascension of a star 23

scale factor 221
Schwarzschild radius 144
second of arc 65
spectra
    continuous spectrum 76
    line absorption spectrum 79
    line emission spectrum 78
spectroscopic binary 85
speed of electromagnetic waves 51
spiral arms 120
spiral galaxies 125
steady state theory 169, 227
strong nuclear force 176
superclusters 128
supernova 11

telescope 166
thermal radiation 50

ultra-violet radiation 49
Universal Constant of Gravitation 41

virtual photons 177
voids 128

weak nuclear force 178
white dwarf 74

X radiation 49